The Painted Desert and Petrified Forest

(Evidence for a Catastrophe)

Dr John G Leslie

Deluge Press

All Photographs, Text, and Drawings done by John G Leslie.

ISBN: 0-9979076-1-4
ISBN-13: 978-0-9979076-1-2

Deluge Press
Gallup, New Mexico

DEDICATION

To my wife Barbara, mother of our children and my soul mate for over 39 years. To my children Peter, Anna, and Teresa. Also to Nancy, Greg, and Jennifer, my siblings. All who appreciate the beauty of God's creation. Finally, to my parents, who taught me to love God and the wonders of His creation.

"The man who finds a wife finds a treasure and receives favor from the Lord." Proverbs 18:22

"Children are a gift from the Lord; they are a reward from Him." Psalm 127:3

"For the time will come when all the earth will be filled, as the waters fill the sea, with an awareness of the glory of the Lord." Habakkuk 2:14

"Truth springs up from the earth, and righteousness smiles down from heaven." Psalm 85:11

CONTENTS

ACKNOWLEDGMENTS

A special thanks to Dr. Andrew Snelling for his comments on geology and who has been a strong advocate for God's creation and the Noah Flood as a real True Narrative Representation.

Thanks to Jim Brown for his friendship and willingness to review this book.

INTRODUCTION

The Current Views of the National Park Service

Petrified Forest National Park (PFNP) comprises both the Painted Desert, and the Petrified Forest. It covers more than 100 square miles[1] of what is called the Chinle Formation, which is assigned to what has been designated the Triassic Age. It is composed of various layers of red, brown, purple, and gray/white clayey-sandstone. The usual scenario to explain its physical properties state that this area was once located in more Mediterranean latitude with high upland forests and meandering streams in low grasslands. Volcanic activity resulted in occasional local floods and catastrophic deposits of trees, plants, and organisms. Many of the trees were thought to have been dead prior to fossilization and simply fell into the streams-thus having no roots or limbs. Biologic life forms, when found in "death assemblages" are assumed to have occurred during flooding of the local streams. During this time inland lakes and seas were supposed to have contributed as well to the strata by ingression and egression. Movements of the continents (plate tectonics) are used to explain its present latitude. The ages of the strata are thought to be about 205-227 million years ago, thus having formed over at least 22 million years.

This concept of the process of gradual prolonged accumulation of the earth's surface is called **uniformitarian** geology - "as things have been so they are." Over 600,000 people per year hear this story through film, books and the park rangers.

Another View

The physical geology of the park reveals much information that is markedly different to the above story. The flood plains are not just local, but cover a minimum of 150,000 square miles within the uplifted Colorado Plateau region of Utah, Colorado, Arizona, and New Mexico, with a probable extension of somewhat similar strata into Newark, New Jersey, Europe and South America. The strata are up to 1500-1800 feet thick. There is also evidence for continuous sequential deposition of the Chinle strata, compressing into days or weeks what was expected to take millions of years. Many of the fossil trees have evidence of limbs torn off, bark partially or fully removed and root bulbs with no extension of root structures and some evidence of roots having been torn off. Most of the trees appear to have been buried while alive while a rapid heavy over-burden occurred crushing the trees before they had any time to rot. The vertebrate and invertebrate fossil forms are found jumbled, fractured and piled on top of each other in numerous places and levels of strata that have both clayey, ashen, and sandstone properties. A few intact vertebrates reveal shapes consistent with being in the throes of death at the time of burial. These findings occur across the Chinle Formation.

Thus, the evidence can be consistent with a catastrophic flood of regional and possibly worldwide proportions in which there was continual ebbing/flowing of the mixed watery mud/ash flows punctuated by volcanic activity over broad areas of land and massive drainage which then resulted in the present geologic structures. This is called **Catastrophic** geology.

Artistic Renderings of the Contrasting Views of the Formation of the PFNP

Top Figure: Drawing portrays in a limited fashion that the trees were blown down and/or inundated by muddy/sandy laden waters, most still alive at the time of burial. In some areas there are jumbled logs, but in others there are large trees lying in parallel which would more likely be due to vast large moving bodies of water. Life forms are found scattered, broken, and in death assemblages through various strata. Many petrified trees show evidence consistent with rapid burial and heavy compression upon them. They have flattened growth rings in cross section. A few petrified trees have eroded cores that may reflect rotting before burial. And some have arthropod borings. This is not inconsistent with a massive flood.[2] These observations will be viewed and discussed in light of catastrophic deposition.

Bottom Figure: Drawing is a rendition of a painting that has been in the Petrified National Park Museum previously. It shows some living but other dying, rotting trees within a dense forest. The dead ones, mostly Araucaria, fall occasionally into a meandering small river and partly rot. They are then gradually covered by local river sediments. Other murals include two Phytosaur fighting with each other and with blood dripping from the sides. Some reptiles are shown eating lush verdant vegetation. There is one poster that portrays a meteor having hit the earth which supposedly caused a dramatic shift in the environment, and resulted in the death of the so-called Dinosaur Age. Each of these concepts will be challenged in the following pages.

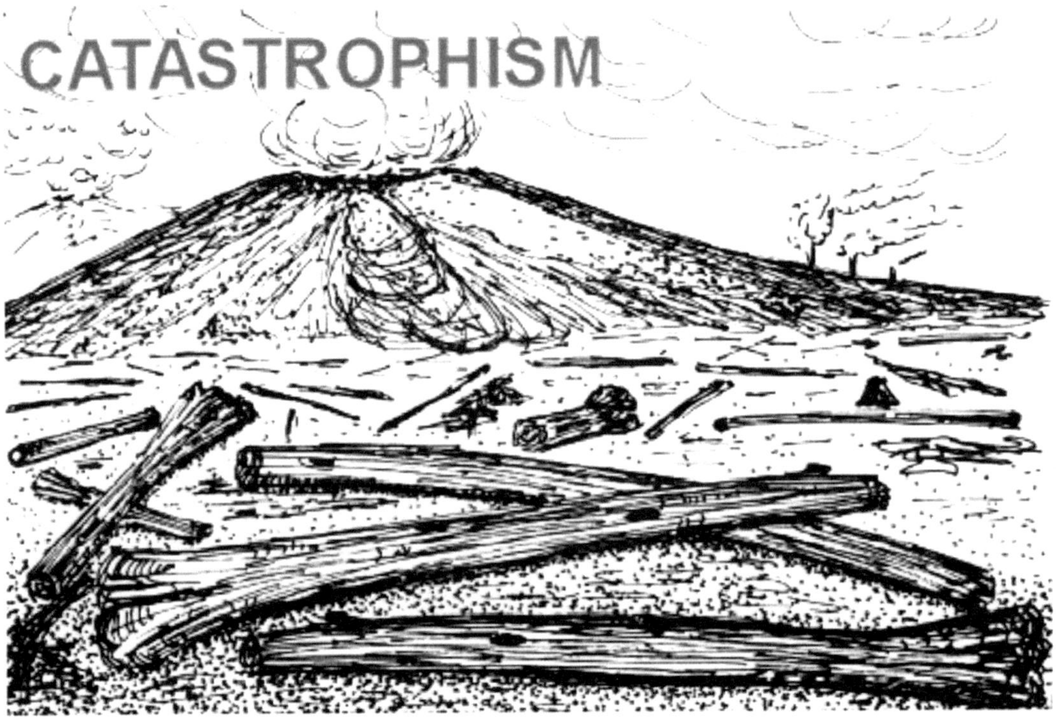
CATASTROPHISM

UNIFORMITARIANISM
JGL

National Park Service Poster from 1935-43 WPA[3]

PART 1 A TOUR THROUGH THE PFNP

Colorado Plateau Uplift

The Colorado Plateau, also known as the Colorado Plateau Province, is a physiographic region of the Intermountain Plateaus, roughly centered on the Four Corners region of the southwestern United States. The province covers an area of 337,000 square km (130,000 square mi) within western Colorado, northwestern New Mexico, southern and eastern Utah, and northern Arizona. About 90% of the area is drained by the Colorado River and its main tributaries: the Green, San Juan, and Little Colorado Rivers. Most of the remainder of the plateau is drained by the Rio Grande and its tributaries.[1]

The Colorado Plateau is composed of various rocks of continental origin, including claystone, sandstone, limestone, siltstone, and conglomerate. Please note that these strata are consistent with having been laid down under watery conditions in most situations. In the southern part of the Colorado Plateau it ranges in thickness from a wedge edge to slightly more than 1,700 feet but is generally more than 1,000 feet thick. The Chinle Formation rests unconformably on the underlying strata. In most of the Colorado Plateau it overlies the Moenkopi Formation. Along the eastern margin of the Colorado Plateaus province, it rests on strata of Permian age. On the flanks of the Uncompahgre and Front Range highlands, it overlies Precambrian igneous and metamorphic rocks. In places it is uncomfortably overlain by younger Jurassic or Cretaceous strata.

The Chinle Formation and related strata are divided into a lower (bentonitic) part and an upper (red-beds) part. The lower part consists principally of variegated bentonitic claystone, clayey siltstone, clayey sandstone, and thin widespread layers of sandstone and conglomerate. It contains an abundant fossil fauna and flora, including pelecypods, gastropods, arthropods, fish, amphibians, reptiles, and plants. The upper part consists principally of reddish horizontally bedded or unstructured siltstone and generally minor amounts of limestone, ripple-laminated siltstone and sandstone, limestone pebble conglomerate, and cross-stratified sandstone. Fossils are much scarcer in the upper plateau than in the lower and are mostly pelecypods and gastropods. The contact between these two parts of the formation is gradational and intertonguing and is difficult to locate in some areas. The lithologic differences between the two parts indicate differences in environment of deposition and in source of sediments.[2] Please note that the terminology of the strata has been modified since 1972 till now 2017.

Elevations on the plateau range from 3,000 to 14,000 feet with an average of 5,200 feet. Due to the high range in elevations, the climate ranges from Sonoran Desert to Alpine, however, semiarid conditions prevail.[3]

Please note that the plateau contains surface strata that appear to have been laid down under massive watery conditions. There has been much uplifting and uneven distribution of strata above it and below it. It will be contended that this can be consistent as being a microcosm of a broad-based flood, even worldwide.

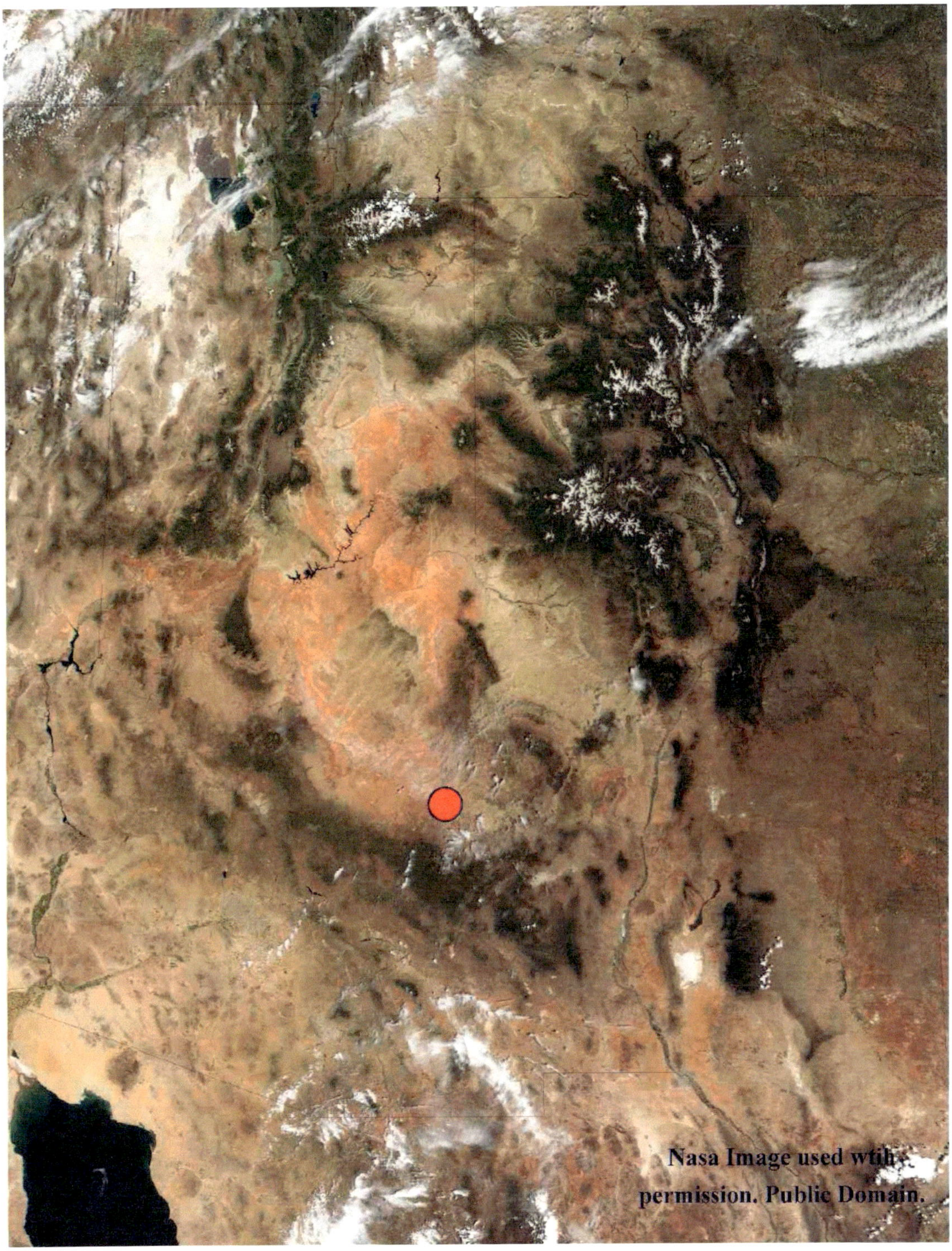

The Colorado Plateau of the Four Corners Area (NM, AZ, UT, and CO juxtapose here). The Chinle designated Triassic is exposed where much of the pink/orange shading is presented. The Petrified Forest National Park resides within this area (red dot added – approximate location).

Amenities in the PFNP

Bookstores

Petrified Forest bookstores are operated by Petrified Forest Museum Association, a non-profit organization providing funding, staff, and supplies through bookstore sales. Your purchases directly benefit the park.[4] However, **I cannot recommend the geology books**.

The bookstores are at the following locations:

- Painted Desert Visitor Center (North End before Tiponi Point area)
- Rainbow Forest Museum & Visitor Center (South End next to the Giant Logs area)

Restrooms

Facilities are found at the north and south end of the Park in the Visitor's Centers, The Painted Desert Inn (midway through the drive) and the Museum.

Painted Desert Inn National Historic Landmark

Located at Kachina Point, 2 miles (3.2 km) from the north entrance at exit #311 off of I-40. The inn once served as a respite for travelers along historic Route 66.[5]

Rock Shops

There are many rock shops found off the Park that have legally obtained petrified wood samples. One such shop is located at the Hopi Travel Center on I-40 just east of Holbrook, with its main shop in St John AZ.[6]

Museum

The museum, at the south end of the Park, has some fossil remains and dioramas. Unfortunately, it is all placed within a long age evolutionary development philosophy. Nonetheless, the fossils are very interesting, both reptile and plant. **I cannot recommend the books**. Just take your own pictures and make a travel log.

Petrified Forest/Painted Desert Map

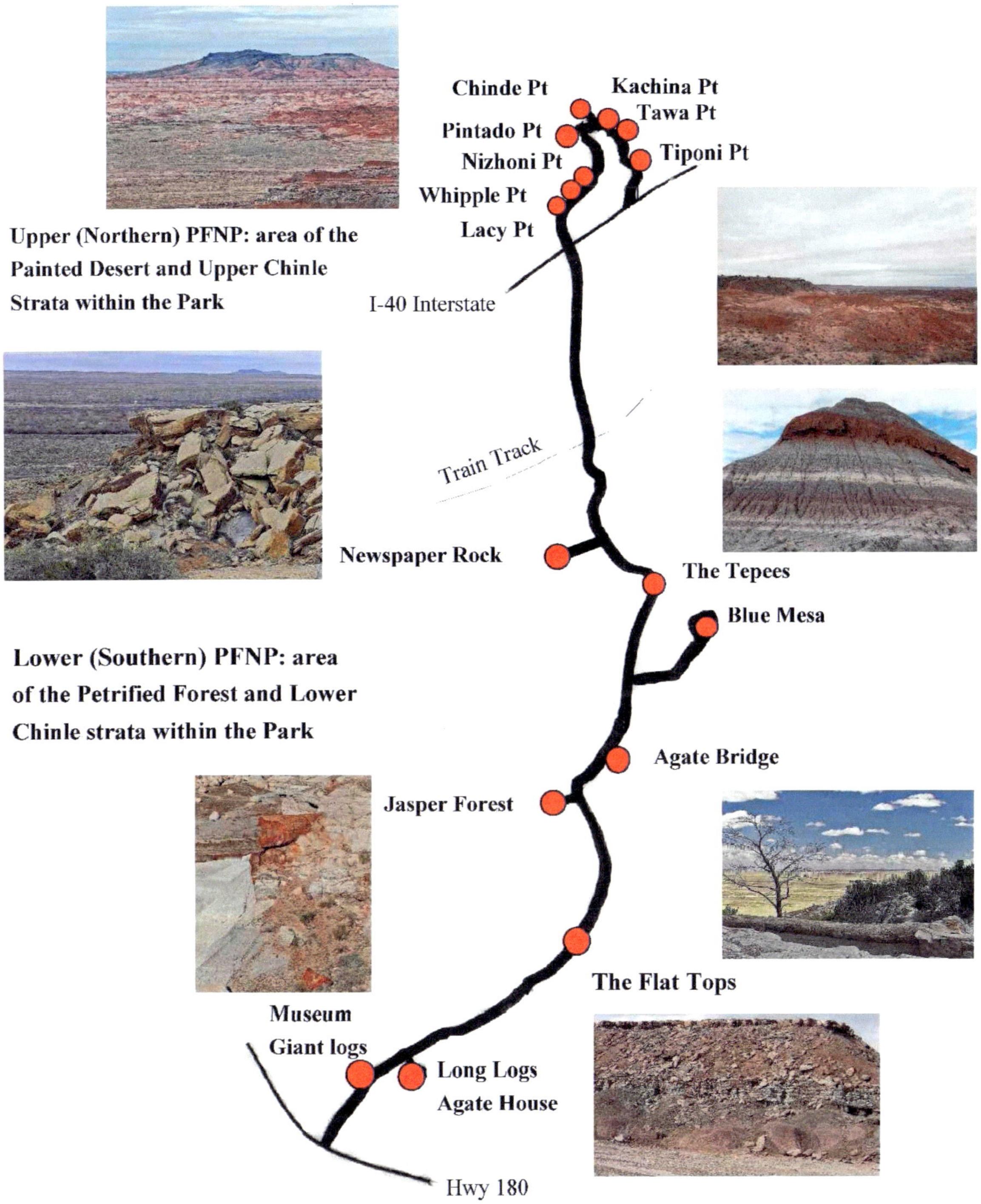

Artwork/photos by J.G. Leslie but some information extracted from the map and writings at the PNFP.

Painted Desert - Tawa Point

The eroding mounded hills seen in this area of the park are composed of clayey-sandstone layers.[7] Iron contributes to the reddish layers (oxidized) and grayish (reduced) layers, whereas manganese to the purplish ones.[8] Whitish gray bands have increased carbonate or silica concentration. While some refer to these layers as growth soil plains, it is more accurate to call them flood plains composed of mixtures of volcanic ash, mud, and scant organic remains.[9] These layers do not demonstrate the horizons seen in growth soils; and there are scant tropical fossil vegetation identified in them (see Geology section). Evidence for forests is lacking. As well, the composition of the strata and the aridity of this area still support only sparse vegetation. Often, the layers are seamless from one uniform layer to another. In other areas there are alternating and inter-digitizing bands between layers that correlates with a process of ebbing and flowing. Yet these same processes are seen over large areas[10], and are best explained by massive flood deposition.[11] Photograph was taken from the Tawa Point lookout, the north area of the drive through the Park.

Owl Rock Formation-Chinde Point

This area "consists of pinkish-orange mudstones mixed with hard, thin layers of limestone. Lenses of selenite gypsum are scattered periodically" as seen in the Owl Rock Member.[12] The Park attributes the selenite to dry lake beds but selenite can be precipitated out of water by other means. (See bibliography on formation of selenite precipitation out of water.)

Bidahochi/Chinle Unconformity-Nizhoni Point

An unconformity, in uniformitarian terms, is where two layers of earth lie juxtaposed where they should not. At this particular point, lava dated at 2 to 4 million years old by certain methods, lies directly on the Chinle strata which has been dated by certain assumptions in radiometric analysis at 230 million years old (See References for excellent article on the problems with radio-dating.). Missing by the usual concepts of geology, are the upper Triassic, Jurassic, and Cretaceous Strata. However, the lack of these layers can be the result of at least two mechanisms, both requiring catastrophism on a scale not seen by present processes. Firstly, it is possible that in this area those strata never existed. All sedimentary strata show large variations in depth and extension, even ones widely dispersed as in the Chinle. Secondly, if these layers were eroded off, it would have involved strata thousands of feet thick and which covered many thousands of square miles. As well, unconformities are also found below the Chinle Formation.[13] When river delta masses of the present age are measured none reveal such volumes of debris having been removed or deposited under current rates of erosion. This one unconformity alone requires results in which either of these processes would need to have happened on a scale not observed today. Yet it is consistent with the ebbing, flowing, and erosive processes of a massive worldwide flood. Photograph from Nizhoni Point lookout, the north/western area outlook.

Painted Desert (From Tawa Point)

Owl Rock Formation (From Chinde Point)

< Pinkish strata

Bidahochi Formation (From Nizhoni Point)

< Lava layer

< Upper Chinle

Newspaper Rock

This lookout is located on the west side of the drive around the Park. It is designated as part of the Blue Mesa Formation that has a sandstone cap of the Lower Sonsela. See Geology section for more of a description of its formation. It has numerous petroglyphs on portions of the sandstone that had already fallen from the cliff face. See Miscellaneous Section for close-up photographs. The petroglyphs are thought to be 600-1100 years old. There are anthropomorphic (human shaped), zoomorphic (animal shaped), and other forms. Some are thought to be related to Hopi images (Katsina) or Zuni (Koko) as representing spirit shapes. Others are geometric and some may represent astronomical related images. It is thought that one marks the summer solstice. The anthropologists of the Park associate the glyphs with Basketmaker (500-650 AD) and Pueblo I, III, and IV (650-1450 AD). It is thought that some Navajo markings occur around 1750-modern.[14]

The Tepees

These geologic formations are part of the Blue Mesa Formation. They have distinct lines of bluish, purplish, and light grey clayey bentonite layers. As the layers are soft they are easily eroded and thus they form a tepee shape. There are also occasional thin bands of sandstone in them.[15] As they are considered part of the Blue Mesa Formation they represent some of the lower layers of strata in the Park. See Geologic Section. Please note that because the clayey matter is much softer that the sandstone there is differential erosion that has occurred. The sandstone will overhang the clayey layers at times.

Newspaper Rock and The Tepees

Newspaper Rock

Tepees

Blue Mesa

This area is considered to be some of the "oldest" strata of the PNFP as it underlies the lower part of the Sonsela Sandstone Member. It is comprised of many grey, blue, purple and green mudstones and occasional areas of sandstone. The Newspaper Sandstone is considered part of this Member per the PNFP. It is located mid-way through the tour. The Park has commented: "The lowest (insert oldest) well-defined rock unit in Petrified Forest National Park is the Blue Mesa Member of the Chinle Formation…streams deposited the strata of the Blue Mesa Member…These rocks represent part of a vast muddy tropical floodplain.[16] There are sandstone beds, the most prominent of which is the Newspaper Rock Sandstone. This unit is best exposed in the Tepees area of the park." As well, reptile fossil forms such as "*Acaenasuchus*[17], *Desmatosuchus*, *Calyptosuchus*, *Hesperosuchus*, *Leptosuchus*, *Smilosuchus,*[18] are found in the strata of the Blue Mesa. It is important that many of the fossil vertebrates are found crushed together in what are called "death assemblages."

It should be noted that the sandstone beds are often conformed on top of the clayey bentonite layers. This could be consistent with continuous deposition of the various strata while still in a wet yet partly dried state, rather than gaps with many years in between each strata deposition. More erosion might be expected if there were large gaps of time between depositions.

Agate Bridge

Most of the petrified wood, scarce plant life, and amphibian/reptile remains of this area are found in or associated with sandstone. The process of agate fossilization often involves replacing each molecule of wood with silica, and with calcium, iron or manganese which contribute the colors. The rate of fossilization is affected by pH, temperature and pressure, as well as the concentration of minerals, especially silica, and the flow of water. Synthetic fossilized wood can be rapidly produced for commercial purposes. (See bibliography on processes of fossilization.)

Many of the trees in the Sonsela Sandstone (a major layer mid-level in the Chinle Formation) are oriented in a northeastern direction indicating significant water flow in that direction. Others, including Agate Bridge, are oriented in a southeastern direction. In some areas the bedding planes of the sandstone indicate that water flows and mud (bentonite) deposition occurred in several directions. This would be consistent with significant turbulence during strata deposition which occurred over a broad area.

Blue Mesa and Agate Bridge

Blue Mesa (above) Agate Bridge (below)

Jasper and Flat Tops Beds

These two beds geologically have been described as the Jasper Forest bed/Rainbow Forest bed. Sandstone and conglomerate, with common cobble-sized clasts of extrabasinal volcanic rocks, chert, and quartzite. These sandstones are usually multi-storied, with the dominant bed form being planar cross-bedding with lesser trough cross bedding and horizontal planar bedding. Reddish and multicolored petrified logs are abundant in these units, as are vertebrate fossils. The Jasper Forest bed and Rainbow Forest bed are names for equivalent units respectively exposed north and south of the Flat Tops in southern PFNP.[19]

There has been some controversy as to the number of layers, extension and nomenclature of the Jasper and Flat Tops beds:

It is curious therefore that Cooley identified the Sonsela sandstone bed and Rainbow Forest sandstone bed as being stratigraphically distinct within Petrified Forest National Park, with the Rainbow Forest sandstone bed occurring slightly lower in the section, near the top of the lower Petrified Forest Member. This convention has been followed by most subsequent workers even though neither unit can be traced continuously across the Flat Tops. Most of these authors claimed to be able to identify the Sonsela sandstone bed southwest of the Flat Tops above the Rainbow Forest sandstone bed, usually as a thinner and finer-grained unit than the thicker and more conglomeratic bed capping Blue Mesa and Agate Mesa north of the Flat Tops. Important exceptions are Roadifer, who claimed that the Rainbow Forest sandstone bed could be identified north of the Flat Tops about 20 feet *above* the Sonsela sandstone bed, and several authors, who suggested that the Sonsela sandstone bed and Rainbow Forest sandstone bed were correlative. This latter possibility was also suggested by Roadifer and Murry although they seem to have favored the interpretation that they are separate units.[20]

It may well be that the ebbing and flowing of continuous flood waters could explain the ambiguities seen in trying to map the strata.

Jasper Forest

Flat Tops

Long Logs/Giant Trees

Even in the park brochures, it is acknowledged that the trees in the Long Logs area represent a log jam; yet directly across from it within 1/2 mile the Giant Trees area does not give that appearance. But almost all trees show that significant abrasive forces have assaulted them before and during deposition. Most trees show limb sockets without limbs, root bulbs without roots, and often the bark is missing. Many trees reveal forces of compression, having been buried while still alive (not rotten) and then flattened with heavy overburden. This is similar to the appearances of trees seen at Mount St. Helen after recent volcanic activity. In this area the trees are also associated with sandstone, especially in the Giant Logs area. As well, petrified logs are found throughout the vast area of the Chinle. Some of the logs are massive: Ten foot diameter and one long log has been measured 115 feet and another greater than 145 feet.[21]

It is significant that only scant amounts of roots and virtually no leaves/limb structures are associated with these logs, nor are they in many other areas of the Chinle.

Uranium halos[22], in which radioactive uranium decays to lead and leaves tracer marks in crystalline material, occur in the logs of the Colorado Plateau region. Dating by this method has given a much younger age for them than by other methods. It also supports the concept of rapid fossilization in contrast to the concept of prolonged periods of fossilization, as the halos would not have been formed or they would have been destroyed.

A USGS paper describes the development of strata in the Upper Triassic as stratigraphic panels of the Chinle Formation across the Uinta Mountains into northeastern Utah and northwestern Colorado near Dinosaur National Monument depicting the sequence of depositional environments in this region. Conglomerates and sandstones at the base of the Chinle represent the initial development of fluvial systems. These fluvial strata contain mostly carbonate intraclasts and very few siliciclastic pebbles, a composition indicating an intrabasinal source for the pebbles. It is likely that the ancestral Rocky Mountains were reduced in elevation or even covered by Chinle deposits at this time. Fluvial systems were succeeded by a large lacustrine system. Aspects of the lacustrine facies, such as the presence of analcime, indicate not only that the lake was at times saline and alkaline, but also that it was, at least in part, contiguous with a large lake system interpreted to have existed in the Upper Triassic.[23]

All of these characteristics are consistent with massive, watery burial of the logs, rapid fossilization and not of a burial in a swamp, local stream, or the periphery of an inland lake/sea area.

Long Logs (upper in B/W)

Giant Logs(lower)

Museum

The museum has dioramas that portray the Chinle late Triassic times as one of inland streams and upland forest inhabited by reptiles and amphibians such as the Placerias, Phytosaur, Metoposaurus, Desmatosuchus, and the Coelophysis. Mollusks including the Unio arizonensis, and plants such as the Calamities are listed. Common trees of the Chinle, including Araucarioxylon (of which a living tree is very similar), Schilderia and Woodworthia are displayed. It is told that all of these conditions and life forms gradually passed away, being replaced with what is now present.

Yet each of the fossils and their geologic conditions reveals a different story of catastrophic proportions. The Placerias, Metoposaurus, Phytosaur, and Coelophysis are usually found in large deposits called death assemblages. Often they are found crushed or fragmented (see photos next page). These findings are interpreted by the PFNP as occurring in watershed areas, but the frequency of such assemblages, the numbers involved, and the fact that often articulated animals are found crushed together argues against this. As well, many similar findings are found across continents and thus support the concept of an initial one land mass some call Pangea.[24] Mixing with other flora and fauna not expected to be found with these animals is also supportive of a catastrophic burial. Phytosaurs, within variation, are also found on most continents, and the Desmasuchus is found spread across Arizona/Texas reflecting a wide distribution of the burial process.

Mollusks[25], which are generally found in the lower layers, are often found in death assemblages fragmented/broken, as well as in concretions (usually in the half shell) consistent with conditions of rapid harsh burial. Many are of the Unionidae form. Another group, the Conchostracans are found just below the Coelophysis bone bed in the Chinle Formation at Ghost Ranch.[26]

Plants such as ferns, similar to plants found in tropical environs and not the present environment have been identified in isolated sandstone formations. The Calamites, a fragile large hollow plant, which has been estimated to be up to 30 ft. in height has been found in Colorado Triassic strata as upright and packed aggregates. It appears as though they had been washed and silted together. Trees, as has been indicated, reveal the trauma of an overflowing mud/ash deposition.

All these observations are supportive of rapid massive burials of plants and animal life occurring over widely dispersed places in the Chinle Formation and with probable transport of some of the animal and plant remains over long distances.[27]

Fossil Life Forms in the PFNP Museum

Biologic Forms

(Sampling and Crushed)

Placerias

Metoposarus

Phytosaur

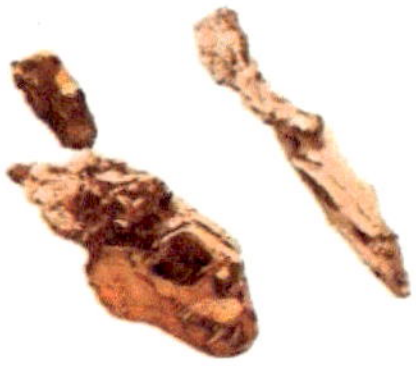

Vancleaver

Crushed clams

Unio clam

(in concretion)

(These clam shells viewed outside PFNP)

Botanical Forms

(Sampling of fern leaves flattened)

Cycad leaf

Fern leaves

Matonia Fern leaf

Royal Fern leaf

(See Ferns from the Chinle Formation [designated Upper Triassic] in the Fort Wingate Area, New Mexico GEOLOGICAL SURVEY PROFESSIONAL PAPER 613-D, 1969, by S.R. Ash in bibliography for further discussion.)

Touring car: a 1932 Studebaker displayed at the entrance to the Petrified Forest National Park.

PART 2 MORE VIEWS TO SEE AROUND THE PFNP

This section presents other information, pictures, and artwork about the Petrified Forest National Park and surrounding area that were not included in Part 1.

Heinrich Balduin Mollhausen's etching of the Whipple Expedition of 1853 going through the Lithodendron Wash. This comes from a poster along the tour route of the PFNP. H. Balduin Mollhausen (1825-1905) was a German writer, traveler, and artist who visited the United States and wrote novels based on his experiences.[1]

The indigenous people of northern New Mexico by Balduin Mollhausen[2]

Painted Desert Inn (National Landmark)

Herbert Lore built the Inn before 1920, but registered it with the AZ Land Office in 1924, during the Homestead Act. He sold American Indian artifacts and gave tours of the area. As well, he had a couple of rooms to rent. His original home was built of petrified wood. In 1936 he sold the property to the Petrified National Monument. The Inn was modified by the CCC (Civilian Conservation Corp) during this period to have an adobe motif. It opened again as an Inn in 1940, but was not used till post WWII. Fred Harvey renewed the business there by hiring Mary Jane Colter and Hopi artist Fred Kaboti to restructure the building. Arrangements were made with the Santa Fe Railroad to bring tourists into the area. It continued till the 1950's. It was inactive again till 1976 when it was placed on the National Registry of historical structures, and was opened on a limited basis.[3]

The Inn before remodeling by the NPS. Building was composed partly of petrified wood.[4]

Skylight panels (above) originally done by the CCC in the 1930's, now refurbished.[5]

Mural Painting (left) done on one wall by Hopi artist Fred Kabotie.

It was commissioned by Mary Jane Colter during redesign of the building in 1947 to 1948.[5]

Puerco River Pueblo in the PFNP

The pueblo is speculated to have been inhabited during the Pueblo IV period from about 1350-1600 AD.[6] The Village on the Rio Puerco (or Puerco Pueblo, for short) is a 100+ room pueblo site located near the Puerco River, a major drainage that bisects the park. The river would have been a reliable source of water for crops. Farming of corn, beans, and squash took place on the floodplains and terraces along the river. At its largest size, around 1300, Puerco Pueblo may have been home to about 200 people. The one-story high village of hand-shaped sandstone blocks was built around a rectangular plaza.[7]

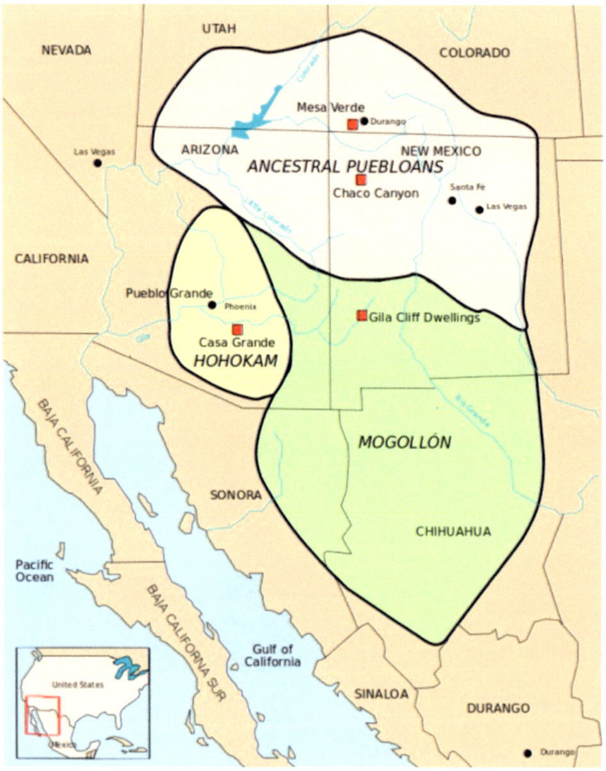

Left: Map of Ancient Pueblo People in the American Southwest and Mexico. During the Pueblo IV period, Four Corners pueblo settlements were abandoned (northern and central portion of the Anasazi region.)[8] Right: Colorado River drainage with tributaries. Pueblos formed along the water shed area.[9]

Puerco River Indian Ruin in the PFNP

The Serpent figure (above) is common among North American Tribes, and among other tribes of the world. However, a human face (next to it) is uncommon. The Polychrome basin is in the Park Museum and is typical of the Pueblo II-III Period from about 550 AD—1300 AD according to the PFNP. The Pueblo is assigned to this time period by the PFNP. The square structure, below ground, is a kiva. According to the park's brochure, there are 1,000 archaeological sites within the Park.

Newspaper Rock and Petroglyphs

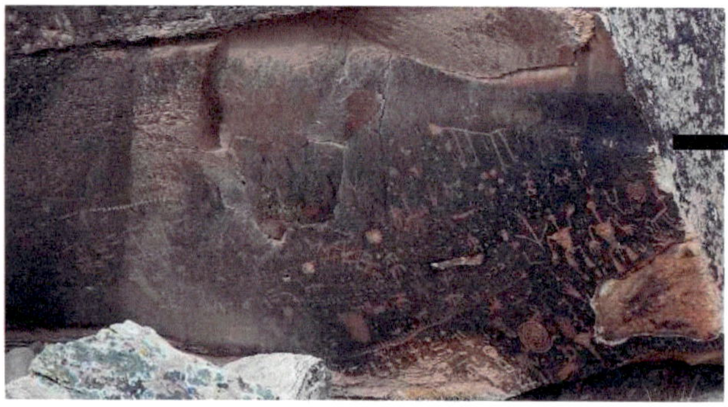

Puerco River AZ-NM State Line

The Puerco River (Spanish: Rio Puerco) flows in northwestern New Mexico and northeastern Arizona, through arid terrain, including the Painted Desert. The intermittent river is the main tributary of the Little Colorado River, which is a tributary of the Colorado River. It drains an area of about 2,654 square miles and is 167 miles long. The river's average discharge is very low, less than 70 cubic feet per second in normal years, because its drainage basin is extremely dry. For most of the year, the Puerco River is a braided wash containing little or no water, although large flash floods can occur in downpours. Although the word Puerco means pig, it also is used to mean dirty or filthy in Spanish. This usage in the southwest United States is better translated as Dirty River or Muddy River due to its high content of silt and mud.[10]

Please note that this river which runs through the PFNP does not have the flow necessary to have caused the erosion seen in the Park. The above photo was taken mid-winter at high flow.

Biological/Botanical Life Forms

In and Near the Painted Desert and Petrified Forest National Park[11]

Animal Forms:

1) Killdeer (Charadrius voiferus) white banded black winged bird

2) Dragonfly (species unknown)

3) Common Raven (Corvus corax)

4) American Pronghorn Antelope (Antilocapra Americana)

5) Painted Lady Butterfly (Vanessa cardui)

6) Collared Lizard (Crotaphytur collaris)

Plant Forms:

1) Shrubs in Winter:

    a-Sage

    b-Crisleaf buckwheat

2) Flowers and flowering shrubs in Spring/Summer:

    a-purple flower- Astragals sp. Milkvetch, locoweed

    b-yellow flower- Senecio flaccidus, thread leaf

    c-white flower- Pushia stansburiana, cliffrose

Biologic and Botanical Life Forms in and near the PFNP

(A sampling)

Weather Fast Facts at the PFNP[12]

1) Snow and rain occur throughout the winter months, from as early as October to as late as March.

2) The monsoon season occurs in July and August, providing the majority of annual rainfall. Mornings often dawn clear, but thunderheads begin to form by late morning.

3) Humidity "The average relative humidity of the area is well below 50%, at times less than 15%...average June relative humidity is 26%, while the average August humidity is 47%.

4) Temperatures in Petrified Forest National Park can vary as much as 40 F between the highs of daytime and the lows of night. Winter temperatures vary from 20 F to 40-50 F, and summer from 60-90 F.

5) Winds can gust from 40-60 mph in late winter to early spring.

Climatic Averages 1948-2001
Petrified Forest National Park

	Average Maximum Temperature (F)	Average Minimum Temperature (F)	Average Monthly Precipitation (inches)
Jan	50	22	0.44
Feb	55	26	0.46
March	62	30	0.91
April	69	36	0.50
May	81	44	0.33
June	90	52	0.26
July	92	61	1.85
Aug	89	60	1.72
Sept	85	53	0.87
Oct	73	41	1.18
Nov	59	34	0.55
Dec	49	21	0.58
Total	-	-	***9.65***

Winter in the Petrified Forest

Clouds Over the Painted Desert Areas: Winter and Summer

The Play of Light in the PFNP

Beauty in the Midst of Corruption: Painted Desert

The Painted Desert

Just north of the Petrified Forest portion of the park, in the early morning with moisture in the air.

The Beauty of Creation Seen in Petrified Wood

(The minerals that produce the colors in petrified wood)

Blue- copper

Orange- iron

Pink- iron

White- quartz

Grey- manganese

Black- carbon or Manganese

Yellow- iron

Types of Petrified Wood Found in the PFNP and Chinle Formation
(Considered Tropical)

Araucarioxylon

Cross-section (crushed) | Cross-section (un-crushed) | Bark or outer surface of tree

(Tropical Conifer)

(< This piece photographed in situ within the PFNP.)

Schilderia adamanica (Tropical Confier)

(< This piece photographed within the PFNP Museum.)

Woodworthia arizonica (Tropical Conifier-per PFNP)

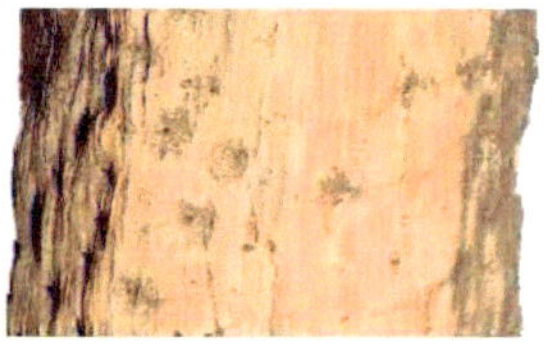

(insert where loss of wood)

Note: Wood (rock) samples come from the three main types of trees found in the Chinle Formation and in different areas of it. Samples (except for the two noted above) are not from within the PFNP. Colors are due to different minerals: iron, copper, manganese, and chromium. Actual sizes vary – photos not to same scale.

View of the Painted Desert looking north from Nizhoni Point.

PART 3 GENERAL GEOLOGY (EVIDENCE FOR A CATASTROPHE)

The Petrified Forest is described by the National Park Service in this way: During the Late Triassic Epoch large river systems flowed northwest through this region to the sea, which was located in what would become Nevada. These rivers deposited thick layers (over 900 feet or 300 meters) of silt, sand, and gravel burying their channels and floodplains. Modern erosional forces have re-exposed these deposits as the colorful badland hills, flat-topped mesas, and sculptured buttes of the Chinle Formation, which makes up a large portion of the Painted Desert region of Arizona. About 60 million years ago this region of Arizona began to rise due to similar forces. The uplift process raised some areas as much as 10,000 feet above sea level. Over millions of years, erosion stripped away many layers of rock. It is thought that erosion stripped the rocks of the Jurassic and Cretaceous Periods before the Chinle rocks were covered by younger rocks of the Bidahochi Formation, creating a break in the rock record—an unconformity.[1] Evolutionary scenarios require long ages and uniformitarian assumptions. Please see Radiometric Dating Methods in the Bibliography section for problems using these methods for the age of strata.

There is much information that the layering of strata in the southwest United States Petrified Forest area and other similar areas were not only due to local rivers and their deltas but due to massive flooding that covered much of the western and possibly continental United States, and that the layers were rapidly sequentially laid down not taking millions of years to occur.

These contrasting views will be discussed within this section.

Description of the Strata in the PFNP: Chinle Formation Composition

Bidahochi Formation: Described as both the result of a lava field flow and "Fine grained fluvial and lacustrine (lake related) sediment such as silt, clay, and sand" and volcanoes "spewed ash and lava over the land." The volcanic "fine grained" ash was deposited in the "lake sediments".[1]

Owl Rock Member: "composed of pinkish mudstones mixed with hard, thin layers of limestone. Lenses of selenite gypsum"[1]

The Petrified Forest Member: "thick sequences of reddish mudstone and brown sandstone layers" Flat tops "reworked volcanistic unit of the Painted Desert"[1]

The Sonsela Member: It has 3 parts in the new nomenclature: "1) the Upper flat tops: thick, cross-bedded sandstone; 2) Jim Camp Wash: blue, grey, and purple mudstones with small sandstone beds; 3) the lower Rainbow Forest bed: cross-bedded sandstone and conglomerate of rounded pebbles and cobbles" Deposit SW to NE over a broad area.[2] The structure of the Sonsela argues for fast flowing water laden with sandy/clayey matter.[3]

The Blue Mesa Member: "grey, blue, purple, and green mudstones with minor sandstone beds"[4]

Shinarump Conglomerate: "The Shinarump Member is a coarse-grained conglomerate sandstone that represents a widespread fluvial channelbelt."[5]

Author's comments: 1) It should be noted that all these strata bear marks of having been laid down under flowing muddy/water mixtures in the midst of active volcanic emission. 2) Gypsum can precipitate out of aqueous (saltwater) solutions and not just drying lake beds.[6] 3) Sand dunes can form under aqueous conditions and cross-bedding supports the idea of alternating motions of the mixtures.[7] 4) Conglomerates with "rounded pebbles" argue for fast flowing aqueous complex mixtures. The Shinarump stratum (which is underneath the Mesa Redondo Member in places) also argues for massive fast flowing deposition to the NW. 5) It is remarkable that similar designated strata occur in the Eastern US (Newark Formation)[8] and Europe (Keuper Formation) which supports wide spread deposition. 6) There are many other strata that appear to have been laid down in aqueous conditions under it and over it (the Triassic) in certain areas.7) Yet, the strata designated as Chinle is not contiguous across the US. Even just east of the Painted Desert in the Midwest US there is no designated Triassic Period.[9] In the Painted Desert area (Southwest US) itself there are no large expanses of the designated Cretaceous Period that over lay it. This supports the idea of massive erosion or that it, the Chinle or Cretaceous and other strata never existed in certain areas. These observations can be consistent with a massive ebbing/flowing of complex muddy mixtures and plate tectonics with mountain building. This is supportive of a catastrophic development of the land area. The strata have the appearance of having been laid down sequentially in a massive watery catastrophe, and not local river flooding, though there may have been local areas of erosion. The Noah flood account describes such an event.

Chinle and Associated Strata Across the Earth

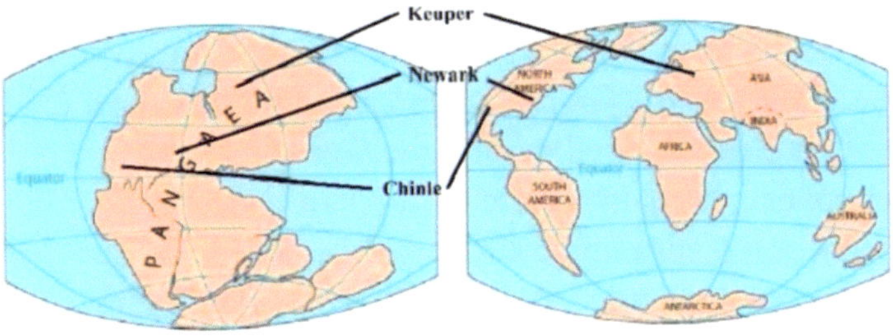

(Designated Permian by the USGS and Createous)
(Map of Pangaea and Suggested Breakup Public Domain and Modifications by JGL)

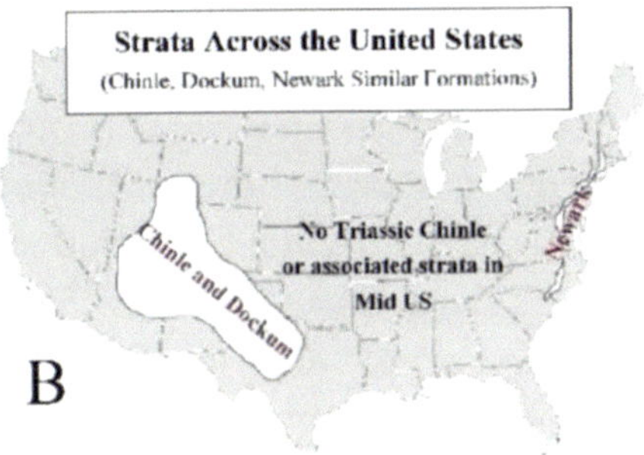

Geologic Column

Please note that there are strata such as the Chinle that show a broad distribution across continents. The nomenclature changes at different sites (but not the composition) and completeness of sequences varies, but there are a few places where most of the geologic column can be found (per discussion with Dr. A. Snelling). However, PFNP nomenclature has changed at least 10 times since the early 1900's. The PFNP strata is exposed on the surface of the Colorado Plateau at present, possibly the result of massive erosion and uplifting. For example: Red Butte has Shinarump strata and it is on the uplift next to the top of the Grand Canyon. For more information see page 63 as well as bibliography under Formation of Strata of the PFNP.

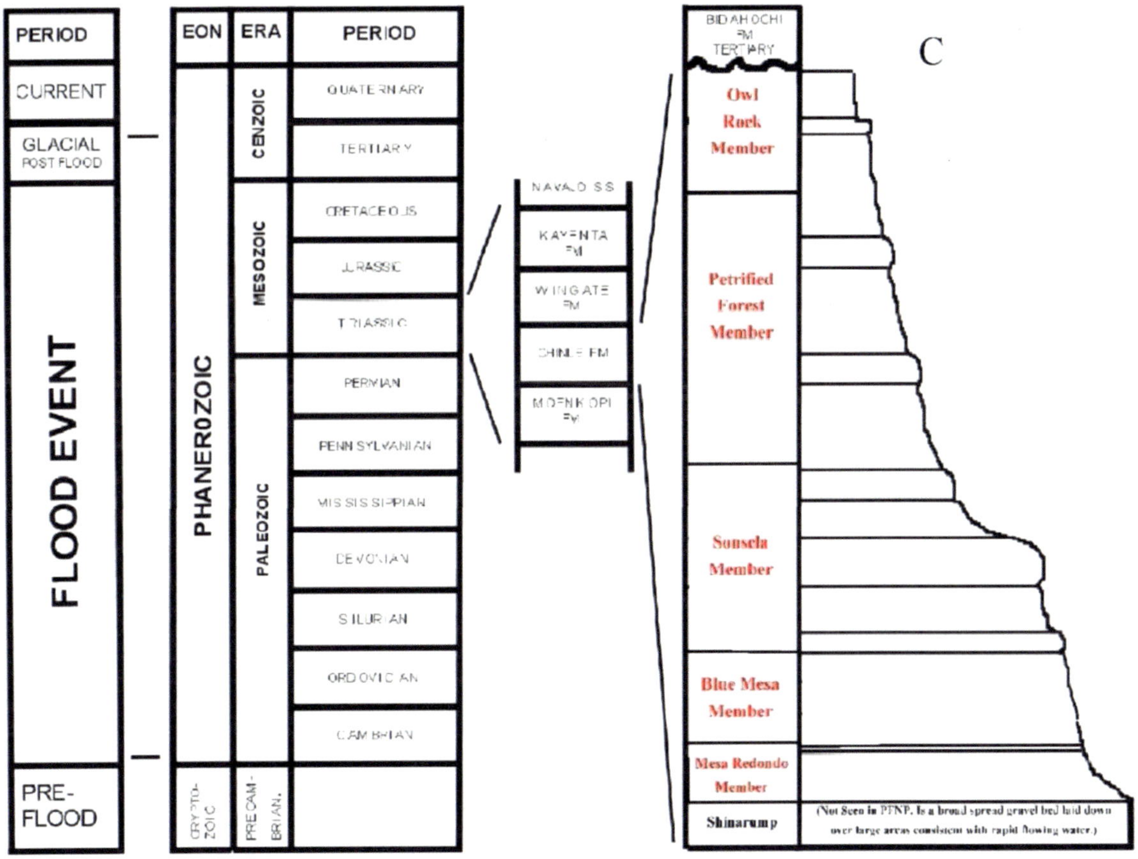

Massive Ebb/Flow of Muddy Ash/Clay or Sand Laden Strata across the United States

The Coconino Sandstone is a good example of massive watery transport of physical earthy debris across a long distance. Dr Whitmore has commented, "The Coconino Sandstone, whose buff-colored sand piles up as thick as 300 feet (90 m) in the Grand Canyon area and 1,000 feet (300 m) in other areas, is one of the most common evidences raised against the Genesis Flood". Young and Stearley argue that the Coconino sand grains are characteristic of modern deserts. Dr. Whitmore analyzed the granules of "sand" in the Coconino layer and found careful study shows the Coconino Sandstone simply does not have the same characteristics as modern desert sand dunes. Evidence that supports this view: 1) slope of strata similar to that found in underwater dunes today of 20 degrees and thus less steep than dry sand dunes today; 2) sand particles were sub-rounded and not rounded like would be found in dry deserts today; 3) Coconino sand grains are not as well-sorted as those in modern dune sands; 4) sand filled cracks in the Hermit Formation, beneath the Coconino Sandstone have the appearance of cracking during an earthquake and filling with "sand ejectites" and that the still soft Coconino sand infiltrated these cracks. Thus, there was no significant time elapsed between the two strata forming; 5) animal tracks best match them being made in watery conditions not dry; 6) "dimples" in the sandstone do not appear to have been made by raindrops in a dry sand; 7) "frosting" of granules probably not by air born erosion but chemical as the granules are very small.[10]

Dr Andrew Snelling has commented (courtesy of Dr. Whitmore), "We can thus imagine how the Flood would deposit the Coconino Sandstone (and its equivalents), which covers an area of one million square miles (2.4 million square kilometers) averages 315 feet (96 meters) thick, and contains a volume of sand conservatively estimated at 60,000 cubic miles or 250,000 cubic kilometers. But where could such an enormous quantity of sand come from? Cross beds within the Coconino dip consistently toward the south, indicating that the sand came from the north. Consequently, this enormous volume of sand would have to have been transported a considerable distance, perhaps at least 200 or 300 miles (320 or 480 kilometers); although some grains come from as far away as the Appalachians. At the current velocities envisaged sand could be transported that distance in a matter of a few days!"[11]

I cite these studies to then refer to the massive flow of the Chinle strata (late Triassic) south east to the northwest[12] and the Cretaceous strata north to the south in the west then north in the east of the US (a cyclical pattern).[13] As well, there is no Triassic stratum noted in the Middle US. These observations are consistent with massive ebbing/flowing/eroding of muddy and volcanic debris.

Examples of Massive Ebb and Flow of Watery Strata Formation[14]: Evidence of Noah's Flood?

Present Day

This topical image reveals the great purturbations that have occurred in North America. Great flooding, faulting and uplifts have occurred across the continent.

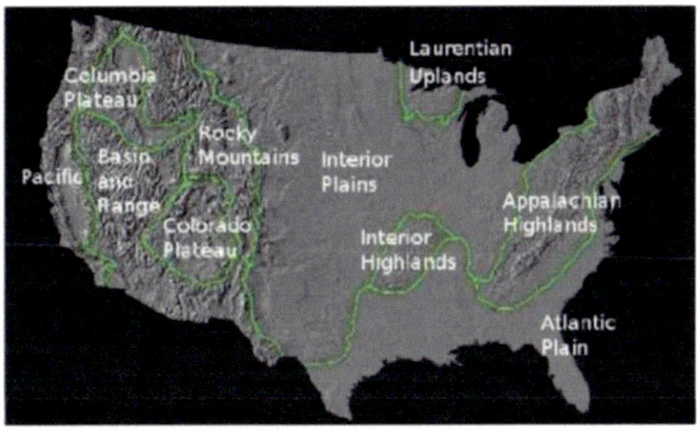

Cretaceous

"Once again, during the Cretaceous Period, record high sea levels flooded the continental interior with shallow seas." Flow is thought to be north to south on the west side and south to north on the east side.

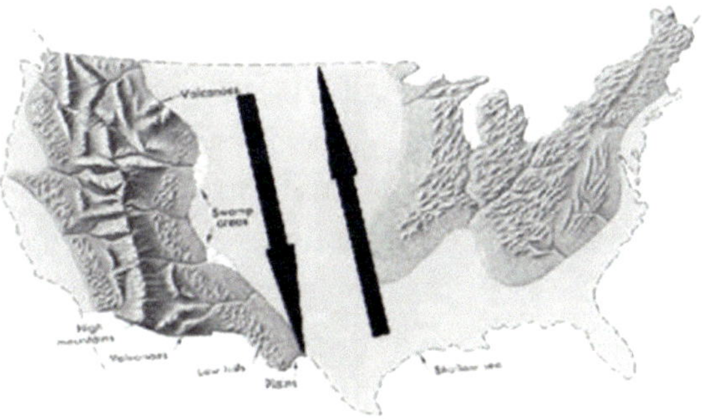

Jurassic

"During part of the Jurassic, rising seas flooded the low-lying areas of the continent. Much of the Interior Plains eventually lay submerged beneath shallow Sundance sea." Similar to Cretaceous in distribution.

Triassic

Was the Triassic strata eroded away or was it never laid down under watery conditions in the Midwest US? Flow was apparently from the southeast to northwest.

Pennsylvanian

"Highland which lay to the east, south, and north supplied much of the sedimentary debris that was spread over the midwestern part of the United States. The Midwest was mainly a low swampy area. Sediment was carried into the region from deltas to the east. From time to time, the level of the sea fluctuated. Swamps were flooded, and forests were destroyed.

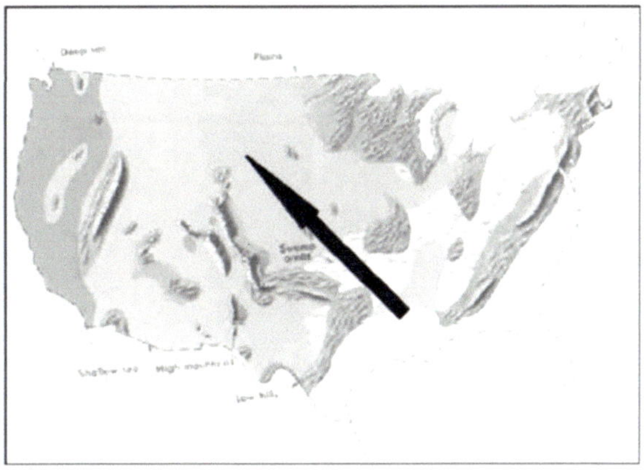

Evidence of a Massive Flow of Water: Long Log Parallel Orientation

There are areas in the PFNP that give clear evidence of a sheets of muddy slurred waters which covered broad areas rapidly. There are several spots, even just driving around the Park, alongside the road where large 40-120 foot logs lie parallel and exposed along their length, often with root bulbs oriented in the same direction(s). These would probably not be logs laid down in a meandering stream. Yet, in modern floods there are examples of non-petrified logs seen that are often perpendicular to the flow of the stream. The roots of the logs may still be in the bank as an anchor. There can be several along a bank. **This is not what is seen in the Park.** One study, of the petrified logs in the PFNP, showed a general orientation of logs with the flow marks in the sand bed below[15], while another in Germany has shown more randomness.[16] Part of the problem is that studies that attempt to compare fossil log and modern log flood deposits use the logs from modern small/medium river floods. As well, logs in larger floods when deposited in basins may have more random orientations.[15] There is evidence, per Dr Snelling, that massive flooding resulted in continuous deposition of strata in the Southwest United States with "extensive, fairly flat erosion surface that occurs above most of the strata units found in the Grand Canyon."[17]

Chinle Formation: Long Log Parallel Orientation in the Upper and Lower Areas

(Support for fast flowing mass of watery ash/clayey mixture)

(Orientation of root bulbs Upper Chinle logs uncertain)

Upper ⇧

Lower ⇩

Destruction and Deposition of Trees during Flooding: Comparison of Chinle Formation and Mount Saint Helens

There is little question that the deposition of the Chinle formation resulted from flowing water on a grand scale. Almost all of the features described in the Chinle are interpreted as features compatible with a broad flood plain. The trees are generally described as having been transported in by flood conditions from source areas outside of the sedimentary basin. In this article they also comment on "signs of water wear" and isolated leaf and twig impressions. They identified 50 species within 40 different genera. Also observed were upright tree stumps which did not have a soil plane below them. They partially excavated two of these trees to the apparent root level. There was no stratigraphic break associated with the base of these trees which could be interpreted as "soil." This suggests that they may be of allochthonous origin (transported to the site of deposition), rather than in position of growth.[18] (Please note that this very concept was later demonstrated at Mt St Helens.[19]) They then presented a model that suggested the trees could have been transported in the sedimentation of tremendous quantities of volcanic ash (of which the Petrified Forest member is largely composed) in a body of water resulting from the Genesis flood. They commented on the ebbing and flowing of the "volcanic mud." Analyses were done on the orientation of trees at eight different locations in the Chinle strata in New Mexico, Utah and Arizona. It was proposed that trees were rafted in from a forest devastated by the flood and became stranded in the volcanic mud as the waters ebbed and flowed in the basin. Subsequent deposition completed the process of emplacement. The orientation of 739 trees was determined in 8 locations. There was a clear orientation of the logs at each spot except one area where the strata where difficult to identify. The preliminary data showed a very strong orientation of logs at most sites. Other studies have used tree orientation to determine the flow of water during flooding in current times. The long axis of the log was oriented to the flow of water.[20]

It should be noted that most current studies involve evaluation of flooding of rivers, with overflow of banks and not true massive flooding; thus many studies are done to assess river damming due to debris. Mt. St. Helens is somewhat of an exception to the rule, though it still has involved localized flooding but on a much larger scale.

Comparison of Trees at Mount St Helens and PFNP

(Evidence Supportive of Catastrophic Deposition of Trees at Both Sites)

< Mt St Helens 15 yrs post event-
(All photos to the left)
PFNP present >
(All photos to the right)

Row 1: Trees without branches or bark

Row 2: Root bulb without roots

Row 3: Ash flow over blown down trees

This photo only from USGS: MSH80_smith_creek_blast_area_channel_incision_09-80.
All others taken by author.

Catastrophic Deposition and Fossilization of Trees

Figure Legend (A): There have been many studies done on the compressibility of raw or dry wood.[21] Most are done for commercial uses including lamination, flooring, ridge and structural purposes. Studies, even in the late 1800's through to the present include the effects of heat, pressure and moisture content. Those dealing with artificial impregnation of the wood with minerals have also included factors such as the pH, solubility of different minerals and pressure.[22,23] It is clear from these studies that a permineralized woody hard material can be rapidly produced synthetically. If one considers that there were incredible pressures, heat, acidic pH (usually) and soluble minerals involved in the depositions and preservations of the organic matter in a massive flood with active volcanoes, then the rapid deposition and fossilization of trees seems reasonable. Fossil trees are found around the world. The forces needed to cause what is seen in the fossil record are not as widespread today. The fossil record can be consistent with a worldwide catastrophe. As well, the compressibility of the fossil wood argues that it was buried rapidly and before it had time to degrade.

Figure Legend (B): 1) Volcanic activity blows down whole forests with most trees oriented in a direction away from the blast. 2) Ash and muddy clays/sands from the volcano pour over the fallen trees. Local rivers may become initially dammed up and then overflow the trees as well. 3) The blown-down trees, stripped of branches, roots, and bark (mostly), are buried and compacted in these volcanic detriments. 4) This creates conditions favorable to rapid fossilization: water/heat/compression, clayey/ash/minerals including silica that cement together gravel/sand and penetrates the organic wood and then precipitates in the cells and hardens. It can also bind together gravel/sand as part of the overall process of fossilization.[24] 5) Once fossilized (which can occur within days as many trees show no signs of degradation) the hardened fossil trees now await erosion of the soft materials around them. 6) As they become more exposed they fragment because of their brittleness. There are numerous studies that have shown how rapid fossilization can occur.[25] See page 60 on the process of Petrification (Fossilization).

Catastrophic Deposition/Compression and Fossilization of Trees

(A)

Examples of Compression and Non-compression

Examples of uncrushed (top) and crushed (lower) fossil trees: Zimbabwe
Lower left and others Southwest U.S.

(B)

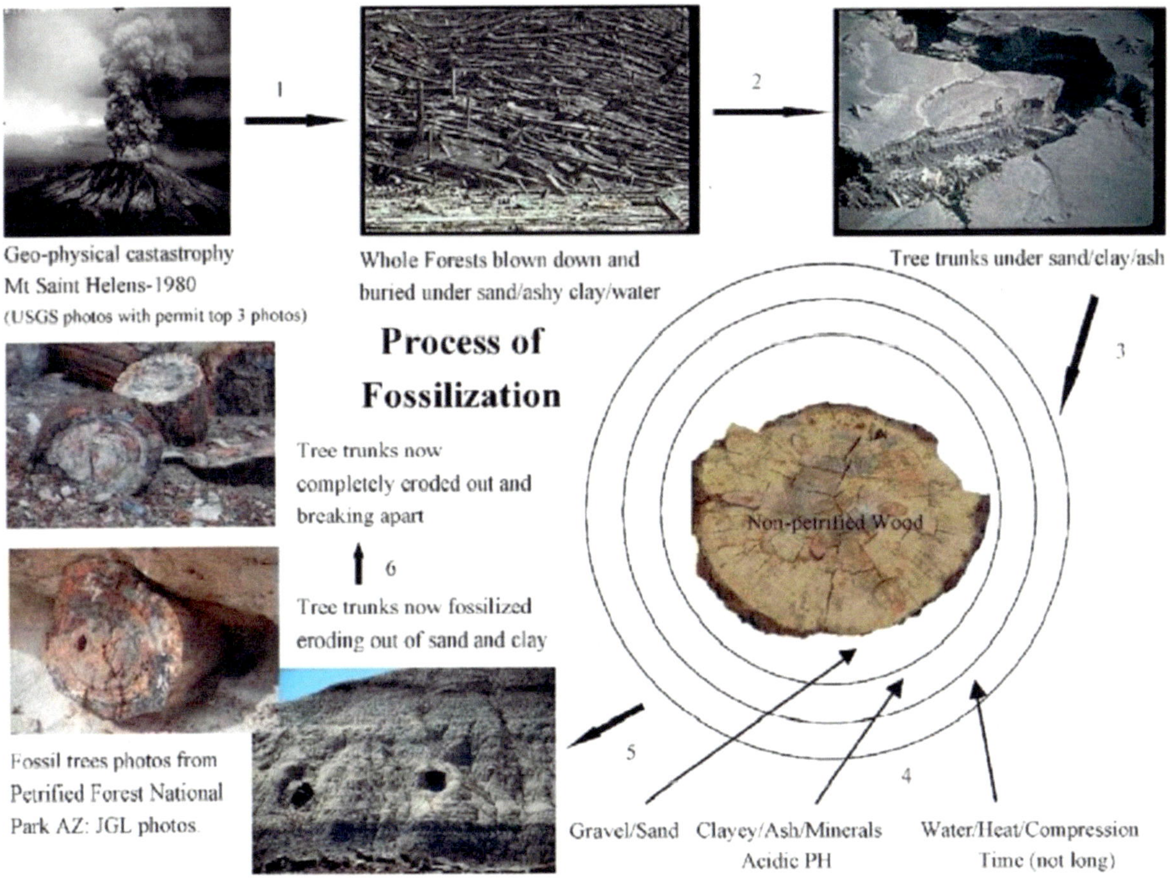

Shilderia: Example of Crushed Petrified Wood

UN-crushed (mostly)

Partially Crushed

Crushed (with different minerals involved in fossilization. Also 2nd compression probably occurred as the striations and rings are maintained but it has been fractured. Samples from outside PFNP.)

Fossilized Root Matter Crushed: In Different Strata

Photographs from Chinle Strata outside of PFNP

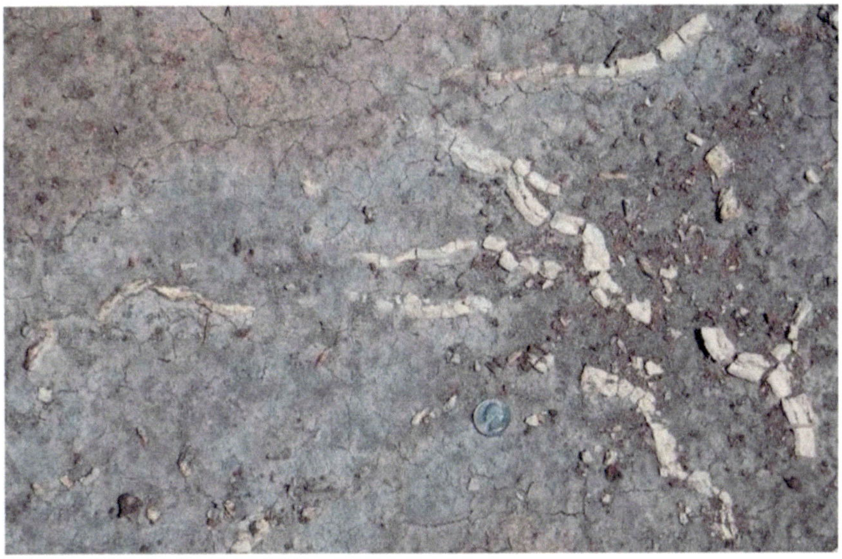

Non-agatized permineralized root matter lying in "mottled" or ash laden Betonite consistent with traumatic burial.

Without thin section analysis it is not possible to confirm these as roots. Some feel similar forms represent arthropod burrows.

Agatized crushed "matted" root or branch matter. These layers were isolated away from the tree trunks. It may include bark matter as well.

side view

Another example of possible multiple layers of matted roots, bark, or leaf material.

Mottled Soil Strata

Soil is weathered, unconsolidated rock debris mixed with at least a little organic matter. The mixture of these two kinds of material is what distinguishes a soil from any other unconsolidated material, such as beach sand, also readily separable into individual grains.[26] Please note that this definition does not imply a plant growth horizon for vegetation only; it can also include flood plains with organic debris. Flood deposit seems apparent in some ancient strata, especially in the PFNP and Chinle Formation, as well as local flooding occurring today. Some examples are below.

Poster illustration in the PFNP Museum.

Mottled strata from near St. John's AZ with apparent fossil disconnected horizontal and upright roots, limbs, or burrows. These can be consistent with flood plain deposition.

Here it appears that there was repeated muddy overlaying stratified layers with some cracking occurring and flow of assumed organic matter into the cracks. This could occur under conditions of heat and rapid sequential layering; note the laminated flow marks into the cracks. These do not appear to be root tracks. Seen near Joseph City AZ.

Here the Chinle clayey strata are directly conformed to the sandstone layers above it. As well, there is a splitting into laminar layers of the greyish presumed organic layer. This can be consistent with fast moving, undulating laminar muddy flow. From St John's AZ.

Polystrate Fossils: Evidence for Rapid Sequential Layering of Water Compounded Geologic Matter

Fairholme 1833

Polystrate fossil upright tree through layers of sandstone. Designated Cretaceous. Gallup NM.

Fossil tree trunk: horizontal but with multiple layers of sandstone covering it. This can be consistent with thin sequential fast flowing watery sand layers that rapidly covered the trunk before it could rot. In PFNP. I have not seen vertical fossil trees in the Chinle Formation.

The **lack** of upright sprouts can be consistent with massive tumbling and rapid burial of the tree trunk. As well, the multiple strata through which the trunk courses reflects the probability that the sequences of sandstone strata were laid down before the tree had time to rot. This is typical of petrified logs in the PFNP.

Regrowth Post Flood Burial: Modern

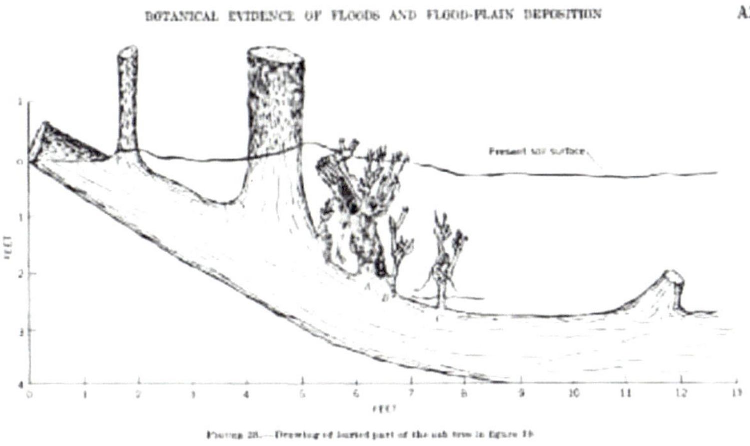

B

The author notes, "The lower part of sprout A grew above ground for 2 years, became buried, and grew for an indeterminate interval beneath the soil surface; later it grew above ground for 3 years and then was again buried and continued to grow below the surface until collected."[27] This sort of pattern of repeated localized river flooding with re-sprouting is **not** seen in the PFNP or the Chinle Formation. Figure from a public domain government publication.

Death Assemblage Fossils: Catastrophic Deposition

Example of the Coelophysis of Ghost Ranch New Mexico

The Coelophysis fossils in the display at the Museum of Natural History of New Mexico (MNHNM) have been partially exposed after the sedimentary stone block they are in was removed and placed in a plaster casting then transported to the MNHNM.

The art work at the Museum has portrayed 4 twisted skeletons not including the isolated bones in this block. Yet, there are reported to be hundreds of fossils found in the Ghost Ranch strata so far. It is considered to be part of the Chinle Formation and dated to the Triassic period because of some other fossils and flora found with it.[28] Many bones are fragmented, with disarticulated joints, but some have articulation. It is felt by those examining the pelvis/femur bones that the organism carried its legs beneath the body and that it made it very mobile. Regarding its death the Museum stated, "Paleontologists are not sure how the Coelophysis in the Ghost Ranch bone bed died. There is no real evidence on the fossils themselves of the cause of death. However, the fact that hundreds of complete skeletons are jammed together in one rock layer suggests a fast and catastrophic death of the dinosaurs. Furthermore, there is **no evidence of weathering or scavenging of the fossil bones, so they must have been buried rapidly after death**." They then comment, "Complete articulation of the dinosaur skeletons is the norm in our block, including sclerotic ossicles in life-position in the orbits. Given this amazing preservation and complete lack of evidence of scavenging or weathering, we believe that the most parsimonious taphonomic interpretation is that **a large congregation of *Coelophysis* were overcome by a sequence of flood events, washed into a topographic low, and immediately buried."**[29] Author's note: They probably died during the catastrophe and not before it. The Museum then states "So, at present, the best inference is that a flash flood killed all the Coelophysis at Ghost Ranch". Author's note: As well, there is **a similar if not the same dinosaur found in Zimbabwe, China, England and South Africa named *Syntarsus*.** It is found in **designated Jurassic** strata, so they ponder how this life form could last 20 million years (see note). Numerous other life forms have also been found buried with the Coelophysis.[30] (Bold lettering added.)

I find it remarkable that they, the paleontologists, who studied this area don't seem to at least consider the idea of at least a large regional flood. This would be reasonable since similar strata are widespread throughout the SW and eastern US.

Vertebrates Found in Death Assemblages in the Chinle Formation

A drawing of the slab of dried clayey bentonite containing the crushed bones of many Coelophysis. Some are articulated and others disarticulated. Over one hundred different Coelophysis have been found in the Chinle Strata of the area of the Ghost Ranch New Mexico.

Exhibit at Museum of Natural History of New Mexico

Concretions

"A **concretion** is a hard, compact mass of matter formed by the precipitation of mineral cement within the spaces between particles, and is found in sedimentary rock or soil. Concretions are often ovoid or spherical in shape, although irregular shapes also occur. The word 'concretion' is derived from the Latin *con* meaning 'together' and *crescere* meaning 'to grow'. Concretions form within layers of sedimentary strata that have already been deposited. They usually form early in the burial history of the sediment, before the rest of the sediment is hardened into rock. This concretionary cement often makes the concretion harder and more resistant to weathering than the host stratum.

There is an important distinction to draw between concretions and nodules. Concretions are formed from mineral precipitation around some kind of nucleus while a nodule is a replacement body."[31]

Studying Fossilization in Concretions

A number of minerals can hold together the concretion, including silica (silicon dioxide), apatite (a type of calcium phosphate) or carbonate minerals — all of which have been found to preserve exceptional fossils.[32] The most successful attempt to replicate fossilization in concretions was by Robert Berner, a paleontologist at Yale University in New Haven, Connecticut, who in the 1960s sealed small fish in jars filled with sea water. The expected chemical changes all took place – increased bicarbonate ions, carbonate ions, ammonia, volatile amines and pH — but there was no deposition of carbonate minerals. Instead, the fish became encased in calcium salts of fatty acids, and natural soaps.[33]

Rapid Petrification

The chemical components used to artificially petrify wood can be found in natural settings around volcanoes and within sedimentary strata. Is it possible then that natural petrification can occur rapidly by these processes? Indeed! Sigleo reported silica deposition rates into blocks of wood in alkaline springs at Yellowstone National Park (USA) of between 0.1 and 4.0 mm/yr.[34]

Invertebrates (Concretions and Fragments) from the Chinle Formation

These fossils were observed on AZ state or BLM Land and not the PFNP; but similar ones are found in the Park. They are found in ash/bentonite as well as sandstone and conglomerates. Many are fractured and crushed. This is consistent with catastrophic burial and re-burial. Please note that living Unio clams are found in fresh water, but prior to the Flood the salinity of any water was probably less. Found in both Upper and Lower Chinle Formations.

Unio clams in concretions half shell above and fractured/collapsed on itself below.

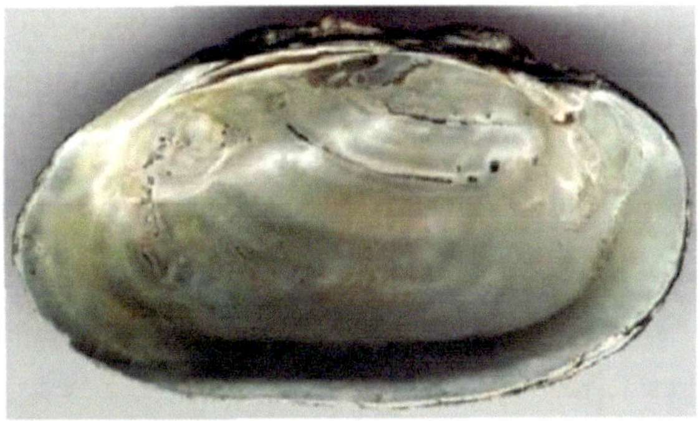

Fresh water unfossilized Unio clam half shell. Similar size and shape to fossil form.[35]

This fossil of a Unio clam probably reflects that it was buried while alive and fossilized rapidly. Its internal organs are petrified and the shell is closed. Fossil clams from outside of PFNP.

These are a mixture of smaller Unio clams and other types with many broken and fragmented.

Please note: Under the right conditions petrification and/or concretion formation can occur in days to months. See Fossilization and Concretion references.

The Process of Petrification (Fossilization and Permineralization)

Yongsoon Shin and colleagues at the Department of Energy lab have converted wood to mineral, achieving in days what it takes nature millions of years to do in such places as the Gingko Petrified Forest, an hour up the Columbia River. There, trees likely felled in a cataclysmic eruption and, buried without oxygen beneath lava, leached out their woody compounds and sponged up the soil's minerals over the eons. Shin's petrified wood journey began in a less dramatic fashion, a few minutes away at Lowe's, Shin's group reports in the current issue of the journal Advanced Materials, in the do-it-yourselfer chain's lumberyard. There they picked up their raw material: pine and poplar boards. Back at PNNL, they gave a 1 centimeter cube of wood a two-day acid bath, soaked it in a silica solution for two more (for best results, repeat this step up to three times), air-dried it, popped it into an argon-filled furnace gradually cranked up to 1,400 degrees centigrade to cook for two hours, then let cool in argon to room temperature.

Presto. Instant petrified wood, the silica taking up permanent residence with the carbon left in the cellulose to form a new silicon carbide, or SiC, ceramic. The material replicates exactly the wood architecture according to Shin.[36]

"The original cellulose structure of the wood acts as a template," said Yongsoon Shin, the PNNL scientist who invented the process. Using a simple chemical process, Shin soaked the wood in acid, then infused it with a source of either titanium or silicon and baked it in an argon-filled furnace. "The innovations here are the use of a wood template for the formation of novel-structured SiC or TiC and the use of wood as the carbon source," said Eric Lund, PNNL commercialization manager. "The newly formed SiC or TiC exactly duplicates the intricate hierarchical cellulose structure of the wood. It also is possible to create the 'negative' of the wood structure by adjusting the initial acid treatment of the wood," Lund said.[37]

It is important to realize that wood and plant material can fossilize rapidly and not necessarily over long periods as has been implied in many writings. (Please see further information on fossilization in the bibliography under processes of fossilization.)

Fossilization: Permineralization

Permineralized wood found in the Lacy Shale Member of the Green River Formation in the Blue Forest WY.

View of the fossil wood partially revealing that it overlaps by bending over another piece of the same sort of wood.

Permineralization is a type of fossilization involving deposits of minerals within the cells of organisms. Water from the ground, lakes, or oceans seeps into the pores of organic tissue and forms a crystal cast with deposited minerals. Crystals begin to form in the porous cell walls. This process continues on the inner surface of the walls until the central cavity of the cell, the lumen, is completely filled. The cell walls themselves remain intact surrounding the crystals.[38]

Mike Viney's article (see bibliography in processes of fossilization) notes that permineralized wood can retain varying amounts of cellulose or lignin. Both of these are the framework upon which silica precipitates during the process of fossilization.

Please note that the wood in these pictures appears to have been buried rapidly while both limbs were pliable and one bent over the other. The cut and polished section also shows some flattening consistent with an overburden being laid on it while it was still an intact pliable piece of wood. Both these observations and the process of permineralization argue for rapid catastrophic burial of fresh wood.

An Example of Massive Erosion and Uplift: Red Butte (at Rim of Grand Canyon)

Red Butte is a butte located in the Kaibab National Forest in Coconino County, Arizona. It is known to the Havasupai nation as Wii'i Gdwiisa, clenched fist mountain, and is regarded as a sacred site. Red Butte is the most notable feature on the Coconino Plateau between the San Francisco volcanic field and the Grand Canyon. Its base is formed of sandstones of the Moenkopi Formation. Above that are strata of the Shinarump Conglomerate; the summit is capped with volcanic rocks.[39] Amazingly, the Moencopi strata lies underneath the Painted Desert Chinle strata just 120 miles to the east. Thus, vast amounts of strata were eroded from around the Red Butte while the faults lifted it up along with forming the Grand Canyon.

Evidence for Massive Erosion: Red Butte

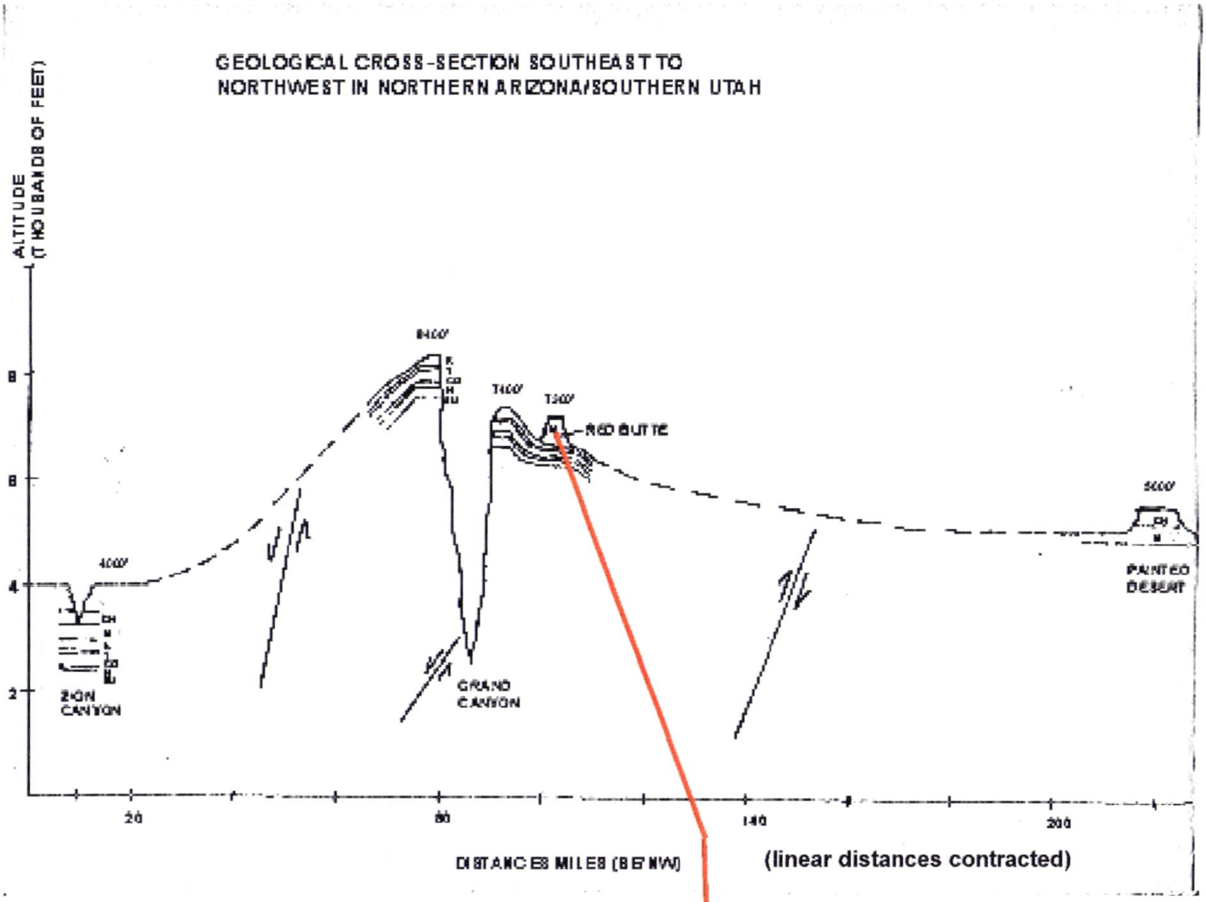

Red Butte

Sandstone/Bentonite Dried and Cracked in Chinle Matrix
Post Modern Flooding with Drainage into the Puerco River Basin
Fossil Crack and Wave Rock in Puerco River Area

These observations support the concept of watery deposition of the bentonite and sand in the Chinle Formation. Although the modern debri was only local, the Chinle Formation is widely dispersed over 4 states. Cracking can occur by short periods of exposure with drying and possibly by compression and heat dessication. Clays are well known for the properties of expansion/contraction. Yet in most layers of the Chinle this is not found and argues for rapid sequential muddy deposition of layers of the sand/bentonite within the Chinle Formation.

Modern flooding in the Ft Wingate area of New Mexico.

Eroded and exposed **fossil wave rock** in Puerco River.

Two months post flood some erosion and break down of cracks while others remain.

Close up of exposed **fossil wave rock** in Puerco River.

Some cracks filling in already.

Exposed **fossil wave rock** in Puerco River.

Drying Experiment with Bentonite and Sand

This experiment helps demonstrate the physical characteristics of expansion and contraction of the bentonite clay of the Chinle Formation. These properties would have an effect in supporting rapid breakup of fossil forms and erosion of strata post watery deposition and drying. Cracking is seen only occasionally in freshly exposed layers, but is prominent in the dried and eroding banks.

5 hours post deposit

Red matrix: mix of sand & clay

Grey matrix: mostly or all clay/ash

(Samples taken from private land.)

120 hours post deposit

Dry: About 2 weeks post deposit

The Painted Desert: Example Of Flooding/Destruction Consistent With Noah's Flood

< Sandstone layers

< Grey and red bands of clay

< Tumbled sandstone blocks due to erosion

< Interface of sandstone and clay layers

Possible mechanism:

Thick layers of moist clay or sand on top of dried clay layers with organic matter.

CONCLUSIONS

Part One

A tour through the Petrified Forest National Park and Painted Desert is a delightful cruise through some beautiful stratified clays and desert plant and animal life. It is also a drop into the past when the glory of Route 66 was in the imagination of many Americans. The tour guide gives a synopsis of this.

Part Two

This section just portrayed some of the human historical aspects of the PFNP, living animal and plant life, and seasons at the park.

Part Three

There are two views regarding the present geologic formations seen in these areas. The first, presented by the Petrified Forest National Park Service, utilizes the concept of uniformitarianism. We have presented an alternative view, catastrophism, in which the geologic observations are viewed to be the result of massive, continuous but ebbing/flowing processes, which could have occurred in worldwide flood conditions. This view is consistent with the Noah Flood Account described in Genesis.

Because each view cannot determine the exact nature or process of the deposition of the Chinle Formation and other strata (as it is a past event not seen by humans), assumptions must be made when deciding which view to accept. These are called presuppositions (see article on religion/philosophy/science in References Creation vs. Evolution):

1) References were made to articles that show quite clearly that radiometric dating is affected by certain assumptions such as preliminary conditions, rates of decay, and other factors that may affect the resulting ratio of the radioactive precursor(s) to the non- radioactive product. Thus, the "age" of strata in absolute years past cannot be determined by radioactive measurements. Long ages are needed if one desires to show some evolutionary tends and to avoid the obvious conclusion of massive flooding. Yet those who choose a young age must assume that rapid sequential strata deposition had occurred.

2) It has been shown in some articles that even super-imposed strata can be of a similar "age". Only the super-position of strata can be determined. And this does vary as one goes across the United States and world. There may be a few places where the entire geological column is represented-per Dr Andrew Snelling. As well, there is also contemporary evidence of rapid sequential strata formation, e.g. Mt St Helens.

3) Up to 73% percent or more of the earth is covered with strata that appear to have been laid down in watery conditions. The sand-dunes of the PFNP can be consistent with water formed dunes. The layers of gypsum could have been laid down in saltwater conditions and not by drying lake beds. So which scenario does one choose? The PFNP chooses to accept the gradual, over eons of time, rising and falling seas, swamps and meandering rivers to produce what is seen. Yet, the same data can be used to support massive or even global flooding. For example, similar strata and life forms, such as the Phytosaur reptile and glossopteris plant are found buried across continents.

4) Many strata contain plants that appear to have grown in a tropical area, as well as forms of amphibians and reptiles. This argues for massive movement of land forms and very different environments before the formation of the strata. It is possible that many life forms could have been carried long distances as well as some shifting of the earth. Mammal forms, though known in the Triassic, have not been reported from the Chinle Formation. But there is a heavy bias towards evolutionary thinking. Sadly, this affects the designation of strata with the assumption that certain fossil types belong in certain "ages". The bias I speak of is clearly demonstrated in the dioramas located at the park museum.

5) There is considerable data that support a massive watery/muddy composite that overwhelmed the plant and animal life now found as fossils in the PFNP. There are large fossil trees that are situated parallel to each other that argue for broad fast flowing waters in which they were carried till buried. These trees are found both in the clayey bentonite and sandstone layers. All, or most, have been stripped of their branches; often with fossilized jagged insert points with the appearance of the limb having been ripped off. Root bulbs, on these large trees, are usually devoid of any extensions of the roots. Yet in many areas masses of root material can be seen compacted together or even as isolated strands in the clayey bentonite. Bark is missing from many trees and those with some bark have the appearance of it being partially abraded off. Only some trees show any decay before fossilization and this is not expected if they simply died and fell into the marsh or stream. There are massive broad strata of fossilized matter that was apparently botanical before being fossilized. This would not be found in a swamp that would degrade much of the matter. As well, these layers are often found encased in sandstone above and below. While root matter may occur in some strata as "mottled" it does not appear to be a true growth soil plain but only jumbled organic matter. Interestingly, there are areas in various strata, or geologic ages, that show polystrate trees: ones that are upright go through several layers and usually do not have any root extension-however, there may be a few exceptions (of partial root extensions).

6) The fossil life forms have, in general, the appearance of having been buried under catastrophic conditions. The Coelophysis death assemblage at Ghost Ranch NM argues for this. The Coelophysis on display at the PFNP Museum shows an arched neck which could be consistent with the reptile drowning and in the agony of death and not just that it died and fell in the stream. There are dinosaur disarticulated bones found in both the clayey bentonite and sandstone, which is more consistent with catastrophic burial. Many of the teeth in the skulls of the Coelophysis appear sharp and in good condition and intact in the skull-except that the skull is crushed. However, there are

many examples of fragmented bones and isolated teeth, especially in the clayey bentonite. This would be expected as the bentonite significantly expands and contracts during drying or moisture.
7) What does one do with unconformities? These are supposed gaps of time, massive at the PFNP of 220 million years old (bentonite layers) and the 80 million years old volcanic Bidahochi layer. One has to make assumptions whether to include a gap of long age or a short age, radiometric dating (without certain assumptions) cannot help one decide.

Overall

It is a wonderful place to visit; just use caution when reading the geology signs. And, it is clear that the PFNP can be viewed as a regional flood zone that may support the concepts of a worldwide flood. The Noah Flood Account in the Bible declares such an event.

"The Deluge" By Gustave Dore[1]

ETERNAL IMPLICATIONS OF A FLOOD IN THE PETRIFIED FOREST

(It is consistent that the flooding seen in the area of the PFNP was part of a historical worldwide event such as the Noah Flood Account)

2 Peter 3:3-7 Most importantly, I want to remind you that in the last days scoffers will come, mocking the truth and following their own desires. They will say, "What happened to the promise that Jesus is coming again? From before the times of our ancestors, everything has remained the same since the world was first created."

They deliberately forget that God made the heavens long ago by the word of his command, and he brought the earth out from the water and surrounded it with water. Then he used the water to destroy the ancient world with a mighty flood. And by the same word, the present heavens and earth have been stored up for fire. They are being kept for the day of judgment, when ungodly people will be destroyed.[2]

There Was a Flood That Covered the Earth

(See Genesis Chapters 6-9)

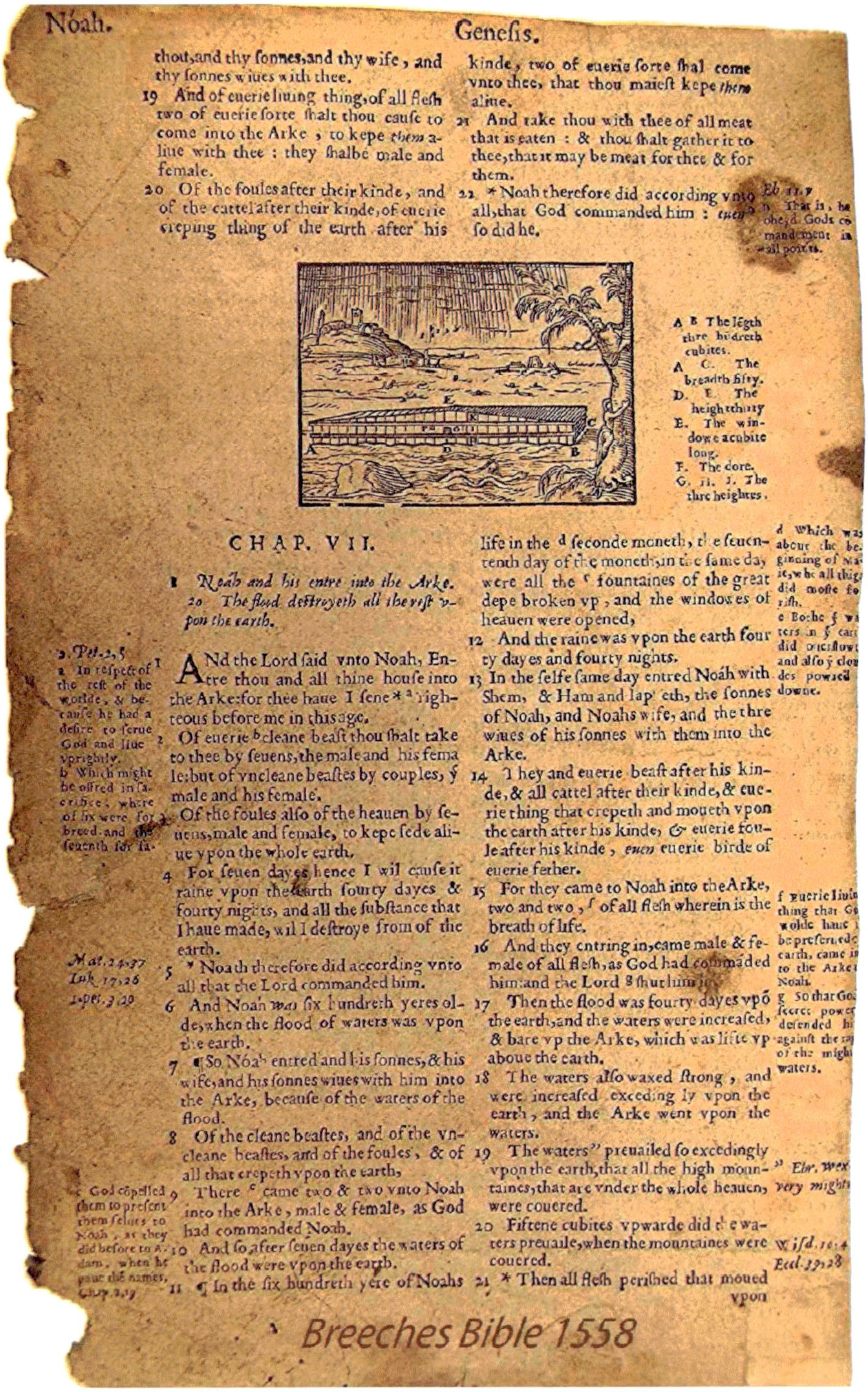

thou,and thy ſonnes,and thy wife, and
thy ſonnes wiues with thee.
19 And of euerie liuing thing,of all fleſh
two of euerie ſorte ſhalt thou cauſe to
come into the Arke , to kepe *them* a-
liue with thee : they ſhalbe male and
female.
20 Of the foules after their kinde, and
of the cattel after their kinde,of euerie
creping thing of the earth after his
kinde, two of euerie ſorte ſhal come
vnto thee, that thou maieſt kepe *them*
aliue.
21 And take thou with thee of all meat
that is eaten : & thou ſhalt gather it to
thee,that it may be meat for thee & for
them.
22 * Noah therefore did according vnto
all,that God commanded him : *euen*[h]
ſo did he.

Eb 11.7
h That is, he obeyd Gods commandement in all points.

A B The lẽgth thre hũdreth cubites.
A C. The breadth fifty.
D. E. The heigh thirty
E. The windowe a cubite long.
F. The dore.
G. H. I. The thre heightes.

CHAP. VII.

1 *Noah and his entre into the Arke.*
20 *The flood destroyeth all the rest vpon the earth.*

1 AND the Lord ſaid vnto Noah, En-
tre thou and all thine houſe into
the Arke:for thee haue I ſene * [a] righ-
teous before me in this age.
2 Of euerie [b] cleane beaſt thou ſhalt take
to thee by ſeuens,the male and his fema
le:but of vncleane beaſtes by couples, ỹ
male and his female.
3 Of the foules alſo of the heauen by ſe-
uens,male and female, to kepe ſede ali-
ue vpon the whole earth.
4 For ſeuen dayes hence I wil cauſe it
raine vpon the earth fourty dayes &
fourty nights, and all the ſubſtance that
I haue made, wil I deſtroye from of the
earth.
5 * Noah therefore did according vnto
all that the Lord commanded him.
6 And Noah *was* ſix hundreth yeres ol-
de,when the flood of waters was vpon
the earth.
7 ¶ So Nóah entred and his ſonnes,& his
wife,and his ſonnes wiues with him into
the Arke, becauſe of the waters of the
flood.
8 Of the cleane beaſtes, and of the vn-
cleane beaſtes, and of the foules, & of
all that crepeth vpon the earth,
9 There [c] came two & two vnto Noah
into the Arke, male & female, as God
had commanded Noah.
10 And ſo after ſeuen dayes the waters of
the flood were vpon the earth.
11 ¶ In the ſix hundreth yere of Noahs
life in the [d] ſeconde moneth, the ſeuen-
tenth day of the moneth,in the ſame day
were all the [e] fountaines of the great
depe broken vp , and the windowes of
heauen were opened,
12 And the raine was vpon the earth four
ty dayes and fourty nights.
13 In the ſelfe ſame day entred Noáh with
Shem, & Ham and Iapheth, the ſonnes
of Noah, and Noahs wife, and the thre
wiues of his ſonnes with them into the
Arke.
14 They and euerie beaſt after his kin-
de, & all cattel after their kinde, & eue-
rie thing that crepeth and moueth vpon
the earth after his kinde, & euerie fou-
le after his kinde , *euen* euerie birde of
euerie fether.
15 For they came to Noah into the Arke,
two and two , [f] of all fleſh wherein is the
breath of life.
16 And they entring in,came male & fe-
male of all fleſh,as God had cõmãded
him:and the Lord [g] ſhut him in.
17 Then the flood was fourty dayes vpõ
the earth,and the waters were increaſed,
& bare vp the Arke, which was lifte vp
aboue the earth.
18 The waters alſo waxed ſtrong , and
were increaſed excedingly vpon the
earth , and the Arke went vpon the
waters.
19 The waters" preuailed ſo excedingly
vpon the earth,that all the high moun-
taines,that are vnder the whole heauen,
were couered.
20 Fiftene cubites vpwarde did the wa-
ters preuaile,when the mountaines were
couered.
21 * Then all fleſh periſhed that moued
vpon

2.Pet.2,5
a In reſpect of the reſt of the worlde, & becauſe he had a deſire to ſerue God and liue vprightly.
b Which might be offred in ſacrifice, where of ſix were for breed and the ſeuenth for ſa-
Mat.24,37
Luk.17,26
1.pet.3,20
c God cõpelled them to preſent them ſelues to Noah , as they did before to Adam, when he gaue thẽ names, Chap.2,19
d Which was about the beginning of Ma[…]
e Bothe ỹ waters in ỹ earth did ouerflowe and alſo ỹ cloudes powred downe.
f Euerie liuing thing that God wolde haue preſerued […] earth, came in to the Arke […] Noah.
g So that God[s] ſecret power defended hi[m] againſt the rage of the mighti[e] waters.
" Ebr. wex[ed] very might[ie]
Wiſd.10.4
Eccl.39,28

Breeches Bible 1558

Genesis[2] (Book of Beginnings)

Genesis 6:5-8 The LORD observed the extent of human wickedness on the earth, and he saw that everything they thought or imagined was consistently and totally evil. So the LORD was sorry he had ever made them and put them on the earth. It broke his heart. And the LORD said, "I will wipe this human race I have created from the face of the earth. Yes, and I will destroy every living thing—all the people, the large animals, the small animals that scurry along the ground, and even the birds of the sky. I am sorry I ever made them." But Noah found favor with the LORD.

Genesis 6:13-14 So God said to Noah, "I have decided to destroy all living creatures, for they have filled the earth with violence. Yes, I will wipe them all out along with the earth! "Build a large boat from cypress wood and waterproof it with tar, inside and out. Then construct decks and stalls throughout its interior.

Genesis 7:11-15 When Noah was 600 years old, on the seventeenth day of the second month, all the underground waters erupted from the earth, and the rain fell in mighty torrents from the sky. The rain continued to fall for forty days and forty nights. That very day Noah had gone into the boat with his wife and his sons—Shem, Ham, and Japheth—and their wives. With them in the boat were pairs of every kind of animal—domestic and wild, large and small—along with birds of every kind. Two by two they came into the boat, representing every living thing that breathes.

Genesis 7:19-22 Finally, the water covered even the highest mountains on the earth, rising more than twenty-two feet above the highest peaks. All the living things on earth died—birds, domestic animals, wild animals, small animals that scurry along the ground, and all the people. Everything that breathed and lived on dry land died.

Genesis 8:1-5 But God remembered Noah and all the wild animals and livestock with him in the boat. He sent a wind to blow across the earth, and the floodwaters began to recede. The underground waters stopped flowing, and the torrential rains from the sky were stopped. So the floodwaters gradually receded from the earth. After 150 days, exactly five months from the time the flood began, the boat came to rest on the mountains of Ararat. Two and a half months later, as the waters continued to go down, other mountain peaks became visible.

Genesis 8:12-17 He waited another seven days and then released the dove again. This time it did not come back. Noah was now 601 years old. On the first day of the new year, ten and a half months after the flood began, the floodwaters had almost dried up from the earth. Noah lifted back the covering of the boat and saw that the surface of the ground was drying. Two more months went by, and at last the earth was dry! Then God said to Noah, "Leave the boat, all of you—you and your wife, and your sons and their wives. Release all the animals—the birds, the livestock, and the small animals that scurry along the ground—so they can be fruitful and multiply throughout the earth."

Genesis 8:20-22 Then Noah built an altar to the LORD, and there he sacrificed as burnt offerings the animals and birds that had been approved for that purpose. And the LORD was pleased with the aroma of the sacrifice and said to himself, "I will never again curse the ground because of the human race, even though everything they think or imagine is bent toward evil from childhood. I will never again destroy all living things. As long as the earth remains, there will be planting and harvest, cold and heat, summer and winter, day and night."

Genesis 9:1 Then God blessed Noah and his sons and told them, "Be fruitful and multiply. Fill the earth.

What Can Fossilized Wood Teach Us?

There are two stories that it can teach us if we will listen:

1. The process of fossilization involves the death of a life form. It happens when a muddy flood pours over the living life form and buries it. The life form dies. In this process it, a tree for example, is stripped of limbs, bark and roots. The trunk is then pressed and flattened. In the right situation of pH, presence of water, minerals dissolved in the water, heat and pressure, the wood will fossilize. The minerals are infused into the cells of the life form, while still organic, and then precipitate out replacing these organic remains.

It is clear that the tree is alive while being buried and crushed; tree rings remain in the deformed and now fossilized stone tree. Over a (short) period of time a precious prized colored gem stone can emerge.

Our lives are the same. We were buried, died, and transformed into beautiful stones (through faith in Jesus Christ). Yet, in contrast to the fossil tree we become "living stones"[3] before God and will stand before Him for eternity.

2) The fossil wood teaches us about God's Holiness, Righteousness, Judgment and Mercy. There was a great Flood which destroyed all the sinful and rebellious people on the face of the earth in Noah's Time, except Noah and his family. The petrified tree reminds of this. So, it is a reminder that God did and will judge sin. Just as there was a worldwide Flood, there will be a worldwide appearance of Jesus Christ at the end of the age. He will judge mankind.

Call upon God while there is time:
Now is the time to repent. Turn to God. Confess your sins to God the Father. Accept that Jesus Christ died for your sins, and was raised from the dead. Proclaim Him as Lord of your life. He will cause your old spiritual nature to die and give you a new one. You will become a beautiful colored "living" stone in His House forever.

Call Upon the Lord

John 17:3 And this is the way to have eternal life—to know you, the only true God, and Jesus Christ, the one you sent to earth.

Matthew 11:28-30 Then Jesus said, "Come to me, all of you who are weary and carry heavy burdens, and I will give you rest. Take my yoke upon you. Let me teach you, because I am humble and gentle at heart, and you will find rest for your souls. For my yoke is easy to bear, and the burden I give you is light."

Romans 10:8-13 In fact, it says, "The message is very close at hand; it is on your lips and in your heart." And that message is the very message about faith that we preach: If you openly declare that Jesus is Lord and believe in your heart that God raised him from the dead, you will be saved. For it is by believing in your heart that you are made right with God, and it is by openly declaring your faith that you are saved. As the Scriptures tell us, "Anyone who trusts in him will never be disgraced." Jew and Gentile are the same in this respect. They have the same Lord, who gives generously to all who call on him. For "Everyone who calls on the name of the LORD will be saved."

A Gap in the Geologic Record
The black basalt that caps the cliffs before

NOTES FROM CHAPTERS

INTRODUCTION

1. New extensions to the Park have occurred in 2014. The addition is 29,093 acres. See NPS in References.
2. It is clear in the scriptures that the earth became corrupted after the fall of Adam. Death entered the earth at that time, and it would be judged in the Flood. Genesis 6:11.
3. Image courtesy of Doug Bouy (image considered public domain, but "used by permission of Ranger Doug's Enterprises").

PART 1: A TOUR THROUGH THE PFNP

1. Colorado Plateau, Wikipedia online 3-2017.
2. Stratigraphy and Origin of the Chinle Formation and Related Upper Triassic Strata in the Colorado Plateau Region, 1972, by J H Stewart, F G Poole and R F Wilson USGS Paper 690, pg. 1.
3. www.nature.nps.gov/geology/education/foos/plateau.
4. https://www.nps.gov/pefo/planyourvisit/visitorcenters.htm It has a private vendor inside.
5. Ibid.
6. C&E Rock Sales 102 S. 2nd , St. Johns, AZ. 85936; Tribal Gift and Rock Shop in the Hopi Travel Plaza, I-40 Exit 292, Holbrook, AZ . (Present as of 4-1-2019)
7. Please see both Reference sections listing articles written by the PFNP and Strata formation.
8. http//scienceviews.com/parks/painteddesert.html
9. J.D. Morris in The Global Flood, ICR, 2005, pg.162.
10. Later it will be shown that the large area of the Southwest US is very similar to the Newark Formation on the East Coast of the US and the Keuper Formation in Germany (see Geology section in bibliography).
11. See references on Strata Formation.
12. See Geology for the PNFP.

13. The Global Flood, pg.119. It reveals massive gaps through geologic formation called unconformities. See bibliography on formation of unconformities.
14. PNFP Bulletin *Messages on Stone*, April 2013.
15. A description of the strata can be seen in any of the sections on geology in PNFP bulletins. I do not agree with their comments regarding the ages of strata (see Geology section this booklet) or the referrals to evolution (see bibliography on Creation/Evolution and Science/Religion/Philosophy).
16. PNFP: Geologic Resources Inventory Report 2010/18, pg. 31.
17. "Vertebrate Fauna; *Stagonolepididae*; *Acaensuchus geoffreyi*," Irmis (2005) pp. 75-77. *Desmosuchus*, *Calyptosuchus*, *Hesperosuchus* and *Leptosuchus* are reviewed in the same text. Most have been identified from a few fragments of fossil. See discussion on "death assemblages" in the geology section.
18. Long, R.A. and Murray, P.A. Late Triassic (Carnian and Norian) Tetrapods from the Southwestern United States, New Mexico Museum of Natural History and Science Bulletin 4 (1995), pg. 254.
19. http://repository.azgs.az.gov/sites/default/files/dlio/files/nid1487/cm-12-a_petrified_forest_report.pdf.
20. http://journals.plos.org/plosone/article?id=10.1371/journal.pone.0009329 (Creative Commons Public Domain Article online 4-13-17 and it was completed with government funds.)
21. https://www.nps.gov/pefo/faqs.htm "Two trees in the Long Logs area of the Rainbow Forest measure 137 and 141 feet long. This indicates that some of the trees may have approached at least 200 feet (61 m) tall when alive."
22. "Creation's Tiny Mystery" by Dr Robert Gentry,1986, publish by Earth Science Associates, Knoxville, Tennessee.
23. https://pubs.usgs.gov/bul/1787w/report.pdf pg W21. Figure 15 also shows a massive lake in the Four Corners Area, and that the Rocky Mountains have not risen in elevation while this deposition was taking place. They place the inundation as local and gradual, but it can be viewed as more rapid and sequential (see geology section).
24. https://en.wikipedia.org/wiki/Pangaea. There is a nice world map that shows the cross-continental distribution of various animal forms, many found in the designated Triassic of the PFNP. I would not accept their age for the strata (see geology section).
25. Mollusks are often found as concretions in the Park and surrounding areas. The process of formation is supportive of rapid catastrophic burial (see geology section). Many are the Unio clam which has living forms. "The Unionidae are a family of freshwater mussels. The range of distribution for this family is world-wide." From Wikipedia. Thus, they are a "living fossil" with worldwide distribution. Pre-Flood probably most or all waters may have been low salinity.
26. In Abstract NM statefossil.org/item/65. Conchostracans are defined as bivalve shrimp found in fresh waters. "Clam shrimp are a taxon of bivalved branchiopod crustaceans that resemble the unrelated bivalved molluscs. They are extant, and known from the fossil record, from at least the Devonian period and perhaps before." They are living fossils, see Wikipedia.
27. This could be one explanation for why tropical plant fossils are found in the Chinle Formation of Arizona.

PART 2: MORE VIEWS TO SEE AROUND THE PFNP

1. 1853 public domain. Wikipedia online 3-2017.
2. 1861 public domain. Wikipedia online 1-2018.
3. Extracted from the PFNP pamphlet "Painted Desert Inn National Historic Landmark".
4. The Inn Image 1937. See NPS online.
5. Both images are NPS photos and include works commissioned by the U.S. Government. Public Domain. Wikipedia online 3-2017.
6. The Pueblo IV Period (AD 1350 to AD 1600) was the fourth period of ancient pueblo life in the American Southwest. At the end of prior Pueblo III Period, Ancestral Puebloans living in the Colorado and Utah regions abandoned their settlements and migrated south to the Pecos River and Rio Grande valleys. As a result, pueblos in those areas saw a significant increase in total population. From: Wikipedia online 4-23-17.
7. https://www.nps.gov/pefo/learn/historyculture/puerco-pueblo.htm
8. Derivative work: Arkyan Map Anasazi, Hohokam and Mogollon cultures-fr.svg: Modifié par historicair 00:26, 9 April 2006 (UTC) - Map Anasazi, Hohokam and Mogollon cultures-fr.svg CC BY-SA 3.0 File:Map Anasazi, Hohokam and Mogollon cultures-en.svg Created: 3 June 2008 online 4-23-17.
9. Map from Wikipedia online 4-23-17 Shannon - Background and river course data from DEMIS Mapserver and The National Map, both public domain. New map of the Colorado River watershed. Endorheic basin in Wyoming fixed, cities added, highlight added, terrain improved. GFDL File:Coloradorivermapnew1.jpg Created: 28 April 2012.
10. Puerco River, Wikipedia online 3-2017.
11. Nps.gov/pefo/learn/wildflower: Please note other plant life are on this site. It is to be regretted that some of the writing regarding the life forms on the PFNP are put within the context of evolution. As well, they have a list of mammals (coyote, fox, mule deer, various skunks, rabbits, mice), amphibians, reptiles, and insects. Photos by John G Leslie.
12. https://www.nps.gov/pefo/learn/nature/weather.htm

PART 3: GENERAL GEOLOGY (Evidence for a Catastrophe)

1. https://www.nps.gov/pefo/planyourvisit/upload/Geology-and-the-Painted-Desert-Site-Bulletin-sb-2013.pdf.
2. *Stratigraphy and Origin of the Chinle Formation.* Geologic Survey Professional Paper. 690, 1972, pg. 40 Fig 20.
3. *Sediment Constraints on the Evolution of the Cordilleran Arc: New Insights from the Sonsela Member.* E. V. Howell, R. C. Blakey in Geol. Soc. Am. 2013, pg. 1 "…the Sonsela Member serves as the most reliable indicator of short-term fluctuations in arc dynamics of the poorly preserved and disjointed Mesozoic portion of the Cordilleran magmatic arc." Online 11-2017.
4. www.nps.gov/pefo/learn/nature/geologicformation

5. *Capitol Reef National Park and Surrounding Areas,* http://sed.utah.edu/Chinle.htm, Front page. Update 2010 online 11-8-17. Some authorities do not place the Shinarump within the Chinle Formation anymore.
6. http://www.swim-site.nl/pdf/swim18/swim18_045.pdf
7. https://answersingenesis.org/geology/grand-canyon-facts/startling-evidence-for-noahs-flood/
8. *The World of Coelophysis* ed. Fisher 1981 describes the Newark Formation. See more info in bibliography.
9. The designated Cretaceous overlaid the designated Permian in much of the mid-west US. See Strata references including USGS Our Changing Continent. But the authors put the observations in the context of local seas rather than a massive flood. Yet, the data could be placed within the context of one massive flood that ebbed and flowed.
10. https://answersingenesis.org/geology/grand-canyon/coconino-sandstone-most-powerful-argument-against-flood/
11. https://answersingenesis.org/geology/grand-canyon-facts/startling-evidence-for-noahs-flood/
12. https://www.nature.nps.gov/geology/parks/nava/geol_history.cfm
13. http://ftp.ems.psu.edu/data/pub/.snapshot/filer3(0135102259)_sm_filer2_ems1.2445/geosc/sling/PDFs_of_pubs/SlingKeen1999.pdf
14. Images from USGS public domain, except topo labels by Hiker 395 on Wikipedia 3-20-17 public domain. Quotes are from https://geomaps.wr.usgs.gov/parks/province/intplain.html; and https://pubs.usgs.gov/gip/continents/. Map without labels from Wikipedia for Triassic. Flow arrows added. Flow information for the Pennsylvanian was also taken from http://earth.geology.yalc.edu-ajs/1993/06.1993.01Garven.pdf.
15. "**There is little question that the deposition of the Chinle formation resulted from flowing water on a grand scale. Almost all of the features described in the Chinle are interpreted as features compatible with a broad flood plain. The trees are generally described as having been transported in by flood conditions from source areas outside of the sedimentary basin. Signs of water wear** are frequently described in connection with the logs... Regarding the **Allocanthous origin** of the prostrate logs, there is little dispute. However, very often the logs are interpreted as being borne in rivers and streams to their present position, but other explanations are also possible...This model involves the **sedimentation of tremendous quantities of volcanic ash** (of which the Petrified Forest member is largely composed) in a body of water resulting from the Genesis flood. **Trees rafted in from a forest devastated by the flood become stranded in the volcanic mud as the waters ebbed and flowed into the basin**...For elongate objects such as tree trunks this means that they will **tend to orient parallel with the prevailing current** Thus the general alignment of logs should represent the average direction of the depositing current. **Uniform orientation at each site would indicate flowing water, whereas random orientation would more likely result from logs sinking in a standing body of water**... The orientation of 739 trees has been determined in 8 locations...These **preliminary data show a very strong orientation of logs at most sites**..." Fossil Tree Orientation in the Chinle Formation.V. Chadwick and L.R.Brand ORIGINS Magazine 1974, pg. 22-28. (Bold lettering added.)

16. "...when individual log orientations are compared with their underlying bed forms, logs may be oriented parallel, sub perpendicular, or perpendicular to the bed form. There is no statistical preference for any particular orientation in the data set. Comparison is made with an actualistic data set from the Lassa Distributary of the Rajang River delta where a similar trend is documented, and with previously published experimental and field data. This comparison indicates that there is no statistical preferred orientation of wood clasts in fluvial systems..." The Relationship Between Bed form and Log Orientation in a Paleogene Fluvial Channel, Weißelster Basin, Germany: Implications for the Use of Coarse Woody Debris for Paleocurrent Analysis R. A. Gastaldo PALAIOS, 2004, V. 19, p. 587–597. How this would change in a massive, not river, situation is unclear.
17. Snelling, A. *Earth's Catastrophic Past*, 2009, pg. 592-596. Some strata in the Grand Canyon are similar to those seen in the PFNP.
18. Fossil Tree Orientation in the Chinle Formation,1974, A.V. Chadwick and L. Brand in *Origins* 1(1): 22-28.
19. Mt St. Helens and Catastrophe,1986, S.A. Austin, *Acts and Facts* 15(7). See bibliography for more information.
20. Study on Flow at River Confluence in River, Costal and Estuaries Morphology,2005, K Mascijin and Y Shimizue RCEM. Hokkaido Univ. of Technology Japan.
21. See bibliography on process of compressibility of wood article by Seborg.
22. See article by Arvidsson on process of compressibility of wood in bibliography.
23. See Kamske, 2008, in process of compressibility of wood in bibliography. He discusses the factors involved in compression/impregnation of wood.
24. It should be noted that in current studies on the impact of burial of trees that they rapidly deteriorate within a few years. Thus, the compression and fossilization would have to occur rapidly before they rotted. See Gholz article in bibliography under process of compressibility of wood.
25. See bibliography on processes of fossilization. There are modern techniques for creating "artificial fossilized wood" within days.
26. Principles of Geology, Gilluly et. al., 4th edition, 1975, pg. 209.
27. Botanical Evidence of Floods and Flood Plain Deposits, Robert Sigafoos, 1964, Geological Survey Professional Paper 485 A.
28. See images from the PFNP Museum. Other fauna include water born reptile Phytosaur and fish coelacanths. Bottom photo is with permission from Museum of Natural History of NM.
29. See www.nmstatefossil.org/item/44 "Taphonomic implications of the microstratigraphy and fauna of a single block from the Late Triassic Whitaker (*Coelophysis*) Quarry (Rock Point Formation: Chinle Group), north-central New Mexico"
30. See www.nmstatefossil.org item/15 "Dig Deeper."
31. From Wikipedia on Concretions online 8-9-17.
32. Victoria, McCoy. 2013. Patterns in Palaeontology: Exceptional Preservation of Fossils in Concretions, Palaeontology Online, Volume 3, Article 7, 1-14.
33. Berner, R. A. 1968. Calcium carbonate concretions formed by the decomposition of organic matter. *Science* 159, 195–197; and online now https://www.ncbi.nlm.nih.gov/pubmed/5634908.
34. Quoted in A. Snelling, 1995, Instant Petrified Wood, https://answersingenesis.org/fossils/how-are-fossils-formed/instant-petrified-wood/

35. Margaritifera margaritifera, double valved specimen, length 108mm, inner side; Teisnach, Germany. 1964. CC BY-SA 3.0. File: Margaritifera margaritifera-binnen1.jpg by Tom Meijer Wikipedia 2-24-17 Fresh water pearl mussel. Only half is shown of bivalve. From Germany, not AZ.
36. *Petrified Wood In Days* https://phys.org/news/2005-01-petrified-wood-days.html#jCp
37. *Instant Petrified Wood Yields Super Ceramics* http://www.pnnl.gov/news/release.aspx?id=1
38. Permineralization on Wikipedia online 6-15-17.
39. Red Butte, Wikipedia online 8-30-17.

ETERNAL IMPLICATIONS OF A FLOOD IN THE PETRIFIED FOREST

1. "The Deluge" by Gustave Dore, January 6, 1832-January 23, 1883.
2. The Holy Bible, New Living Translation, copyright © 1996, 2004, 2007, 2013, 2015 by Tyndale House Foundation. Used by permission of Tyndale House Publishers, Inc., Carol Stream, Illinois 60818. All rights reserved.
3. 1 Peter 2:4-5.

ABOUT THE AUTHOR

1. From http://creationwiki.org/John_Leslie

BIBLIOGRAPHY WITH REMARKS

(Subjects in alphabetical order)

Calcium Carbonate Nodule Formation Within a Watery Environment

Gypsum precipitation/dissolution During Seawater Intrusion N. boluda-botella*, V. gomis-yagües, F. ruiz-beviá and Mª. D. saquete-ferrándiz Department of Chemical Engineering, University of Alicante. Apdo. 99. E-03080 Alicante. Spain. E-mail: Nuria.boluda@ua.es.
Abstract: "The precipitation of gypsum during the intrusion of seawater in a coastal aquifer has been shown as a possible cause of the decrease in sulphate concentration with respect to the conservative freshwater/seawater mixing observed in many field studies."

Coelophysis Distribution and Variation (worldwide in some Triassic and Jurassic designated strata)

A new coelurosaurian dinosaur from the Forest Sandstone of Rhodesia, A. Raath, 1969, Amoldia 4:28.
"A new coelurosaurian saurischian dinosaur is described from the Forest Sandstone (Upper Triassic) of **Rhodesia**. It **compares closely with the American Upper Triassic Coelurosaur *Coelophysis***, but differences in the pelvis and tarsus indicate generic separation of the two forms. Its discovery establishes for the first time the presence of coelurosaurian populations in the **Upper Triassic deposits of Africa, from which the later known Jurassic forms may have developed**." Comment by the NMS Museum "The article that first describes and names *Syntarsus*, a dinosaur from Africa similar to (and considered by some to be another species of) *Coelophysis*." online 7-25-17 Author's comment: 1) speciation is very debatable and subjective. Do the pelvic and tarsus structural differences represent simple variation within a "kind" or a possible evolutionary divergence? Measurements cannot directly address this issue. 2) Evolutionists are troubled by finding similar dinosaur bones in the upper designated Triassic and lower Jurassic and thus assume evolutionary connectedness. In their terms, if the bones are similar it represents a prolonged stasis for the Coelophysis.

Coelurosaur Bone Casts from the Connecticut Valley Triassic, E. H. Colbert and D. Baird, 1958, American Novitates, 1901, pg. 1-11.
"This dinosaur (*Podokesaurus*) shows, in spite of certain obvious differences, various striking resemblances to *Coelophysis*." They then comment on variation and classification, "Figure 2 shows that the several specimens of Coelophysis, ranging from rather small individuals to animals that may regarded as sexually mature adults, form a fairly uniform regression series, as might be expected." They comment on *Podokesaurus* and wonder if the estimated variability of the tibia length allows this creature to be associated with *Coelophysis*. (pg. 8) "...**it is our opinion that the probabilities favor a generic identity between the Portland (Portland Conn. Newark Formation) specimen and the genus *Coelophysis* of the Chinle formation of New Mexico and adjacent states.** There is some temptation to give this fossil a new name in

spite of its fragmentary nature, because of the rarity of dinosaur bones from the Connecticut Valley sediments, but we feel that the conservative course is to designate it as Coelophysis sp." **This reflects that a somewhat arbitrary approach can occur in naming and dividing fossil forms into various groups-species, genus, order etc.** But to their credit they refrain from doing this and allow for some variability within the bones they have (even though fragmentary). They do fall into a certain degree of **circular reasoning** (see section on circular reasoning) in that they discuss that the Portland specimen is in designated "Rhaetic equivalent" strata and the Chinle specimen in a "Keuper equivalent" which is thought to be below the "Rhaetic". From this they conclude, "**Designation of the Connecticut specimen as Coelophysis sp. Implies an extension for the vertical range** (stasis) **of this genus beyond the limits now given it, or a shifting of the Newark group in Connecticut to a position lower than that indicated in the new correlation.**"

Concretions (experimental formation)

Patterns in Palaeontology: Exceptional Preservation of Fossils in Concretions; Victoria McCoy, Palaeontology Online Volume 3, Article 7, 1-14.

'The most successful attempt to replicate fossilization in concretions was by Robert Berner, a palaeontologist at Yale University in New Haven, Connecticut, who in the 1960s sealed small fish in jars filled with sea water. The expected chemical changes all took place – **increased bicarbonate ions, carbonate ions, ammonia, volatile amines and pH** — but there was no deposition of carbonate minerals. Instead, the **fish became encased in calcium salts of fatty acids, and natural soaps. Of course, part of the problem in comparing experimental results with ancient concretions is that even if the experiments exactly replicate the process, the experimental result is the product of less than 1 year of the process,** whereas the concretions can be millions of years old. There is no guarantee that this extra time did not cause significant changes to the developing concretion. Berner suggested this, with the idea that his calcium soap concretions are the precursors of a more typical rocky carbonate concretion." Note this author and Dr Berner struggled with the observation that concretions could form rapidly-years instead of millions of years. His original article: Berner, R. A. 1968. Calcium carbonate concretions formed by the decomposition of organic matter. *Science* **159,** 195–197.

Rapid rock: Unexpected application for hard-rock recipe, Tas Walker, 2002, *Creation* **24**(2):38–39. http://creation.com/rapid-rock.

"The **new invention is simple to use. All that is needed is to spray two solutions onto the porous sand, soil or rock. The water-based solutions seep into the material, replacing the existing pore fluid.** Alternatively, the solutions can be injected into the material. Because the solutions flow easily, like water, the sediment is quickly penetrated. And being entirely non-toxic, the solutions do not pose a health or environmental hazard. **Once inside the pores, the chemicals react to form calcite crystals on the surface of each grain of sediment. The calcite cements the grains together and gives the sediment rock-like strength**. The speed of the reaction can be controlled **from one to seven days** to allow the solution to penetrate into the sediment as far as desired. Because the cement only covers the grain surface, the spaces between the grains remain open. Thus the porosity of the rock is only slightly reduced and the flow of groundwater is not obstructed. This means that the solutions can be applied a number of times to the same sediment and continue to penetrate the pore spaces, adding extra cement each time. The sediment could be converted into almost-solid rock with the pores mostly filled, but this would take many applications and a few months to achieve. Ordinary water will not soften calcite, so the cement bonds should remain strong indefinitely." Note: this article is cited to demonstrate that consolidation of soft earthy material can be rapidly cemented together.

Creation/Evolution and Science/Religion/Philosophy

Religion, Philosophy, Science and How each Interfaces in the Noah Flood Account, 2013, John G Leslie, online creation.com/religion-science-philosophy-interface-noah-flood.

In this article it is discussed how the fields of religion, philosophy and religion interface. All concepts regarding the past formation of the earth involve all three fields.

Formation of Strata in the PNFP and Elsewhere

These reviews on the evaluation of the PFNP strata and associated strata in other parts of the southwest US and elsewhere reveal it is acknowledged that the strata were developed by the ebbing and flowing of watery muddy composites. These articles try to portray that the strata were due to local meandering streams that occasionally flooded. The same data can be placed within regional and even continental (and possible worldwide) flooding.

Chapter W Sedimentology and Depositional History of the Upper Triassic Chinle Formation in the Uinta, Piceance, and Eagle Basins, Northwestern Colorado and Northeastern Utah By Russell F. Dubiel https://pubs.usgs.gov/bul/1787w/report.pdf W10-11.

"**Floodplain Deposits** A large part of the upper Chinle Formation is composed of dark-reddish-brown to reddish-orange**, massive to horizontally stratified, thin- to thick-bedded mudstone, siltstone, and fine-grained sandstone.** Locally, the flat-lying strata can be traced laterally into the lateral-accretion deposits described in the preceding section of this report. Individual beds are laterally extensive cut banks eroded during lateral migration of the sinuous fluvial system; cracks indicate many episodes of channel-overbank flooding and subsequent desiccation." (Please note that desiccation with cracking can occur within hours in still wet bentonite-which the author has observed.)" The horizontally stratified strata contain indurated, isolated or **coalesced carbonate nodules** and several types of **rhizoliths**, or plant-root trace fossils, and invertebrate trace fossils. The lateral relation of these horizontally stratified beds to fluvial point-bar deposits, the early formation and induration of carbonate nodules in the beds, and the association of these nodules with alteration haloes interpreted below as plant-root trace fossils **indicate that this lithofacies represents floodplain deposits lateral to Chinle meandering fluvial systems**." (bold added)

Stratigraphy and Origin Chinle Formation Upper Triassic Strata in Colorado Plateau Region 1972 by JH Stewart, FG Poole, and R Wilson https://pubs.usgs.gov/pp/0690/report.pdf Page 1-2 parts of the abstract quoted below. **Please note it describes water deposition of all strata.**

"The Chinle Formation of Late Triassic age is composed of various rocks of continental origin, including claystone, sandstone, limestone, siltstone, and conglomerate. It extends over most of the Colorado Plateaus p1·ovince, and related strata are recognized in adjoining regions…it ranges in thickness from a wedge edge to slightly more than 1,700 feet but is **generally more than 1,000 feet thick**…The Chinle Formation rests unconformably on the underlying strata. In most of the Colorado Plateau it overlies the Moenkopi Formation, of Early and Middle (?) Triassic age. Along the eastern margin of the Colorado Plateaus province, it rests on strata of Permian age and locally on rocks of older Paleozoic age. On the flanks of the Uncompahgre and Front Range highlands, it overlies Precambrian igneous and metamorphic rocks. The Chinle Formation is conformably, but locally unconformably, overlain for the most part by the Glen Canyon group of Triassic and Jurassic age, but in some places it is unconformably overlain by younger Jurassic or Cretaceous strata.

The Chinle Formation and related strata are **divided into a lower (bentonitic) part and an upper (red-beds) part**. The lower part consists principally of variegated bentonitic claystone, clayey siltstone, clayey sandstone, and thin widespread layers of sandstone and conglomerate. It contains an abundant fossil fauna and flora, including pelecypods, gastropods, arthropods, fish, amphibians, reptiles, and plants. The upper part consists principally of reddish horizontally bedded or structureless siltstone and generally minor amounts of limestone, ripple-laminated siltstone and sandstone, limestone pebble conglomerate, and cross-stratified sandstone. Fossils are much scarcer in the upper part than in the lower and are mostly pelecypods and gastropods. The contact between these two parts of the formation is gradational and intertonguing and is difficult to locate in some areas. The lithologic differences between the two parts indicate differences in environment of deposition and in source of sediments.

The **lower part of the Chinle Formation** is considered to be a continental **deposit laid down in streams and lakes and on flood plains**; this interpretation is based largely on the types of **sedimentary structures and fossils**. Thin widespread sandstone and conglomerate units, such as the Shinarump and Moss Back Members, are probably point-bar deposits produced by the lateral migration of meandering streams. Stream directions, as indicated by the orientation of crossstrata, were mostly north to northwest, predominantly a volcanic terrain The **upper part of the Chinle Formation** may be predominantly **a lake deposit**, as indicated by the fine texture and even bedding of the strata and by the type of fossils. Cross-stratified sandstone layers, interpreted to be **stream deposits**." (bold added)

Depositional Environment of the Upper Triassic Chinle Formation In Eastern San Juan Basin and Vicinity, NM…Trace Fossils and Mollusks…Chama Basin…Lacustrine Origin U.S.G.S. *Bulletin 1808-B-D 1989.* (Note: title shortened but contains 3 separate articles. Only the first article by Dubiel will be reviewed.)

Page B 10 "Red mudstones of the Salitral Shale Tongue exhibit ubiquitous and **abundant endogenic trace fossils** that are preserved in full relief. The trace fossils are **filled with mudstone** having the same composition as the surrounding matrix. The **traces are sinuous and have no preferred orientation**…Isolated sandstone beds in the Salitral Shale Tongue are interpreted as **fluvial deposits** based on their fine to medium grain size and their lenticular geometry. Laterally equivalent purple- and white-mottled **siltstones and mudstones** are interpreted **as levee and overbank deposits of the fluvial systems**. The fine-**grained mudstones represent quiet-water deposits** laid down in small marshes or **lacustrine basins** adjacent to **the fluvial systems**. The superimposed mottles are thought to represent **gleyed paleosols formed on the levee and fluvial units as a result of fluctuating water tables** in a manner similar to that previously described for the Agua Zarca **gleyed paleosols**. The red, massive mudstones at the top of the Salitral Shale Tongue are interpreted as **overbank floodplain deposits. The red floodplain deposits** are also paleosols, but the lack of large mottles…" B 13 "The splay deposits, in turn, **grade laterally into floodplain mudstones** with superimposed paleosol horizons. The sandstones are characterized by an abundance of low angle, complexly crosscutting lateral accretion bedding, **numerous splay deposits**, and **floodplain mudstones that completely encase sandstone beds**. The sandstones probably were **deposited by high-sinuosity streams that carried an abundant suspended-sediment load and probably were subject to numerous avulsion events.** The presence of **fossil riparian phytosaurs and perennialfluvial unionids within the fluvial crevasse-splay strata provide independent complementary evidence for low-gradient, perennial Petrified Forest fluvial systems**." Please note the frequency of the terms fluvial and lacustrine, mudstones and floodplain. As well I do not agree with the concept of Paleosols as ancient soils, as most could be viewed as flood-plains with jumbled fossil plant matter.

Synthesis and revision of groups within the **Newark Supergroup**, *eastern North America* Robert E. Weems *U.S. Geological Survey, M.S. 926A, Reston, Virginia 22092* Paul E. Olsen *Lamont-Doherty Earth Observatory of Columbia University, Palisades, New York 10964*

Western Interior Seaway. Wikipedia online 3-2016.

"The Western Interior Seaway was a large inland sea that existed during the mid to late Cretaceous period splitting the continent North America into two landmasses."

Our Changing Continent, 2-2015, USGS.

The map on page 7 shows a large inland sea covering the Midwest US, but including Texas, the Southern states, and Florida during the Late Cretaceous period. On page 6 they comment: "Shallow, warm inland seas covered large portions of the Southern and Western United States…large portions of California were under water, and in the eastern California, Nevada, Arizona, Idaho, eastern Oregon, Washington, and Alberta." https://pubs.usgs.gov/gip/continents/

Generalized Geologic Map of Kansas http://wwwkgs.ku.edu/General/geolSheetMap.html

This map demonstrates no designated Triassic strata occurs in Kansas.

See PFNP Publications section below- PFNP *Geological Resources Inventory Report.* In this report on pg. 8 the comment occurs:

"The stratigraphic nomenclature at Petrified Forest has been undergoing revisions since the early 1990's. Most recently, investigators have reevaluated the Sonsela Member of the Chinle Formation. Over the years, investigators have separated, combined, translated, and renamed portions of the Chinle Formation."

There is the acknowledgment that the **Sonsela mudstone and sandy components** and that it, the strata of the PFNP and Chinle Formation have been continually revised since the 1950's. Heckert and Lucas 2002 list 10 nomenclatures.

Sedimentary Rock Wikipedia:

"The sedimentary rock cover of the continents of the Earth's crust is extensive (**73% of the Earth's current land surface**), but the total contribution of sedimentary rocks is estimated to be only 8% of the total volume of the crust. Sedimentary rocks are only a thin veneer over a crust consisting mainly of igneous and metamorphic rocks."

http://scienceviews.com/parks/painteddesert.html, 2005 online 3-2017. This site discusses the process of color banding of the PNFP. However, they look at as seasons of drying and flooding. Oxidation dry and red and reduction wet (anoxic) and grey due to iron.

"The presence of iron in the crystal structures of the clay minerals infuses an exceptionally important additional facet into their importance because its oxidation state can be rather easily modified in situ and such changes evoke profound differences in the surface chemical and structural behavior of the clay. Examples of clay properties that are greatly affected by changes in iron oxidation state are swelling in water, cation exchange capacity, cation fixation capacity, surface area, clay-organic interactions, surface pH, reduction potential, ability to transform chlorinated organic compounds..."

Revised Upper Triassic Stratigraphy of the petrified Forest National Park, Arizona, U.S.A., 2002, A. B. Heckert and S.G. Lucas. Originally in the New Mexico Natural History and Science Bulletin, No. 21 https://libres.uncg.edu/ir/asu/f/Heckert_A_2002_21_Revised_Upper_Triassic.pdf online 3-2017

In this article the authors list **ten different nomenclatures** for the PNFP: Gregory 1917, Gregory 1950, Cooley 1957 and 1959, Roadifer 1966, Billingsley 1985, Murry 1990, Ash 1987 and 992, Dubiel 1999 and current paper 2002. Their list is bottom to top: Moenkopi, Shinarump/mottled strata, Bluewater Creek, Blue Mesa, Sonsela, Painted Desert, Owl Rock, and Bidahochi.

All layers are consistent with having been laid down in watery conditions: **siltstone, mudstone, sandstone, conglomerate, carbonate**, and what they call "**pedoturbated**". In Wikipedia the definition of "bio-turbinates" is: "sedimentary geology processes relating to changes in the nature of water-borne alluvial sediments and *in situ* soil deposits over time". Please note that "pedo or bioturbinates" do not necessarily represent soil horizons as many are conformed with the layer above. They often have the appearance of representing root masses laid down in a flood plain.

Earth's Catastrophic Past, 2009, Andrew Snelling, Institute for Creation Research, pg 596.

Dr Snelling discusses so called "missing strata" as it relates to the Chinle Formation and other formations. He stated, "Another related observation is that on top of some strata (e.g. the Grand Canyon area) sequences are enormous fairly flat erosion surfaces. An excellent example is the extensive, fairly flat erosion surface that occurs above most of the strata units found in the Grand Canyon, and forms the upper surface of many of the most prominent plateaus of northern Arizona. The Kaibab Limestone is the prominent formation at the Canyon rim and is the most prominent formation across the erosion surface at eh top of the surrounding plateau country. **That other formations were originally deposited on top of the Kaibab Limestone can be seen in two erosionally isolated remnants that consist of the overlying Moenkopi Formation and the Shinarump Conglomerate Member of the Chinle Formation. This physical evidence for extensive removal of most of the units in the strata column above the Kaibab Limestone is best explained as being the result of sheet erosion, by an enormous body of water that uniformly swept across the vast area...evidence is again consistent with the waters of the Genesis Flood** when they retreated, being responsible for this sheet erosion across Arizona and elsewhere." Author's comment: this may be the same process where sheet erosion removed the Triassic off the Midwest strata as there is no Triassic till the Newark Formation on the East Coast of the US.

Stratigraphy of the Chinle and Moenkopi Formations, Navajo and Hopi Indian Reservations Arizona, New Mexico and Utah GSPP 521-B 1969.

In this survey it is noted that the Moencopi formation covers the entire 4 Corners area of the Southwest US, Fig. 1 B6. It is composed of strata consistent with water-based strata, "...very thin bedded to thin bedded siltstone ledge forming beds commonly display **ripple marks** and **casts of salt crystals**. The sandy beds generally **are ripple laminated and pseudo-cross bedded; this bedding is common in the Moencopi**." Pg. B 7. (Author's comment) As well, in many areas there are so-called (as they are assumed to have existed) unconformities. Examples of this would include the following; "In the southwestern part of the reservations, the Moencopi-Permian contact is virtually a flat surface and has a relief of less than 3 feet. Near Holbrook the Kaibab Limestone was stripped off or was never deposited, and Triassic rocks rest on the Coconino Sandstone of Permian age. In the southern part of the Defiance Plateau, the **unconformit**y has a relief of 1-2 feet and is between the Holbrook Member of the Moencopi Formation and the De Chelly Sandstone of the Early Permian age. In the Monument Valley the Hoskinnini Member, the basal unit of the Moenkopi Formation, rests uncomfortably on the De Chelly Sandstone Member and the Organ Rock Tongue of the Cutler Formation." pg. B 6. The assumed ages for the Early Triassic (Moencopi 200 million years old NPS Wupatki handout; Mid- Permian about 275 million years old Coconino Sandstone USGS handout) gives **a gap of 75 million years** with no apparent strata present and where layers are called unconformed**- this is a is a complete assumption**!

Shinarump Conglomerate in The Global Flood by J. D. Morris, Institute for Creation Research, 2012, pg. 112 and 162. "Known for its abundant petrified wood, this extensive **deposit covers about 125,000 square miles and averages 50 feet thick** Dating to the Triassic system, it too requires conditions of erosion and deposition wholly unlike any we currently experience. **Water movement carried along remains of an extensive forest, along with large cobbles. The size of the boulders and cobbles contained within would take rapid water movement on a regional scale to move them**. What process operating today can do this? The deposition of a thin, uniform bed of sand and gravel of such size differs markedly from anything uniformitarian thinking can propose, certainly quite different from the braided streams we know today, even at flood stage. Hydrologists can only postulate rapid sheet flow of graveling-bearing water, depositing its load as the velocity slowed." He goes on to comment that there is supposed "gap" between it and the Moencopi Formation below it yet there is no evidence for this. He goes on to say, "**The series of Moenkopi Formation and Shinarump Conglomerate is uncomformably overlain by shale, muddy sandstone and volcanic mudflows that comprise the Chinle Formation**." The Chinle Formation itself is considered to be from "several hundred to 1,000 feet thick."

Red Butte Arizona Wikipedia online 8-7-17.

"Red Butte is a butte located in the Kaibab National Forest in Coconino County, Arizona. Red Butte is the most notable feature on the Coconino Plateau between the San Francisco volcanic field and the Grand Canyon. **Its base is formed of sandstones of the Moenkopi Formation. Above that are strata of the Shinarump Conglomerate; the summit is capped with volcanic rocks** because of the slope of the Coconino Plateau, only the uppermost part of the North Rim of the Grand Canyon can be seen." Note: **It is a butte that sits on top of the Coconino Plateau near the Grand Canyon. It is composed of Moencopi and Shinarump (just below the Chinle Formation). Thus, there must have been massive erosion of strata from around it before or as the Plateau was lifting up. (Probably this occurred after the erosion of the Grand Canyon?) This would be consistent with a massive flood event.**

Formation of Sea Floor Strata

Sea Floor Sediment and the Age of the Earth, Larry Vardiman, 1996, Institute for Creation Research.El Cajon, CA.

He comments that assumptions need to be made whether one assumes uniformitarian or catastrophic rates of deposition of sediments in the ocean basins. Using known gamma ^{18}O measurements from sea sediment cores he was able to plot a rapid cooling curve that would be expected post Flood.

The Sands of Time: A Biblical Model of Deep Sea-Floor Sedimentation, Larry Vardiman, 1996. Online at: http://www.icr.org/research/lv/index.html.

In this article he explains the use of measuring Oxygen isotope ratios between O^{16} and O^{18}. "If a catastrophic event such as the Flood occurred, temporary warming of the water immediately after the catastrophe should have occurred and may be detectable. Fossil evidence of water temperature at the time deposition is believed by some researchers to correlate with the ratio of oxygen isotopes of mass 16 and 18." Foraminifera are present in both the fossil record and present day and it is assumed that the temperature of the water at the time of their burial can be inferred by the oxygen isotope ratios. Of course there are problems with this as with any other pre-conditions. But the data can be plotted to support a rapid deposition and cooling post Flood per the author.

Formation of Selenite Precipitation Out of Water

Linked Redox Precipitation of Sulfur and Selenium under Anaerobic Conditions by Sulfate-Reducing Bacterial Biofilms Simon L. Hockin and Geoffrey M. Gadd. http://aem.asm.org/content/69/12/7063.full

"A biofilm-forming strain of sulfate-reducing bacteria (SRB), isolated from a naturally occurring mixed biofilm and identified by 16S rDNA analysis as a strain of *Desulfomicrobium norvegicum*, rapidly removed 200 μM selenite from solution during growth on lactate and sulfate. Elemental selenium and elemental sulfur were precipitated outside SRB cells."

Reductive Precipitation of Selenite in Water Using Sodium Sulfide. S. A. Ahmed (2016 Master's thesis, Texas A & M University. Avail electronically http : //hdl .handle .net /1969,1 /158018.

"Selenium can be removed from wastewaters using biological or chemical/physical processes. However, each of these processes has its own limitation. For example, biological treatment of selenium requires several hours of hydraulic residence time and chemical/physical processes either produces unstable selenium-bearing residuals or concentrated liquid waste byproduct. Reductive precipitation of selenium is an attractive process because it can convert the soluble forms of selenite and selenate to insoluble elemental selenium. This study investigates selenite (Se(IV)) removal from water via reductive precipitation using sodium sulfide at neutral pH. Also, this study investigates the effect of UV light irradiation as an activation method to enhance Se(IV) reduction."

Formation of Unconformities and Supposed Gaps of Time

Grand Canyon Monument to Catastrophe, Steven A Austin, 1994, Instutute for Creation Research. Santee, CA. Regarding the "Great Unconformity", Dr Austin comments on pg. 46 under figure 3.22:

"Detail of the Great Unconformity in Grand Canyon. At this location, Tapeats Sandstone (Cambrian) overlies Dox Sandstone (Precambrian). Evidences of extensive chemical weathering are not obvious. Instead, feldspar-rich pebbles of Tapeats overlie the unaltered surface of Dox Sandstone. Does the surface of erosion represent one-half billion years? We marvel at the fresh appearance of the contact."

Geology from a Catastrophic Viewpoint

The Geology Book, John Morris, 2006, Master Books.
A nice introductory book on the basic concepts of geology and its observations that are consistent with a catastrophic deposition of the strata of the earth.

Earth's Catastrophic Past: Geology, Creation and the Flood. Andrew Snelling 2009, Master Books.
This book takes a more in depth look at geology and the concepts of catastrophic deposition of the surface of the earth.

Mount St. Helens

Mt. St. Helens and Catastrophism by Steven A. Austin PhD (in Geology) www.icr.org/article/mt-st-helens-catastrophism
Dr Austin lists four findings from the Mt St Helens volcanic explosions and their impact on the geology of the region. **1) Rapidly Formed Stratification**: "Up to 400 feet thickness of strata have formed since 1980 deposits accumulated from primarily air blast, landslide, waves on the lake, pyroclastic flows mudflows, air fall, and stream water. A deposit accumulated in less than one day on June 12, 1980, is 25 feet thick and contains any thin laminae and beds." He comments that within conventional geology many seasonal or annual changes would be required for the observations, obviously not true. **2) Rapid Erosion**: He mentions scouring from "steam blast, landslide, water waves, hot pumice ash flows (pyroclastic flows) and mudflows. A mudflow on March 19, 1982, eroded a canyon system up to 140 feet deep in the headwaters of the North Fork of the Toutle River Valley." 3) **Upright Deposited Logs**: He commented on the excoriation of the land with "millions" of prone logs being deposited in the Spirit Lake to make an enormous log mat. Many gradually sank in an upright position. This was documented by underwater sonar. He and others "estimate more than 19,000 upright stumps existed on the floor of the lake in August 1985. The average height of an upright stump is 20 feet…have root masses…some already solidly buried by sedimentation…others have no sediment around their bases." He comments that this might give the possible appearance of "multiple forests which grew on different levels over periods of thousands of years." **4) Peat Layer**: He commented that the rapid accumulation of peat beds at Mt St Helens has the appearance of certain coal beds of the Eastern United States which are assumed to have formed over thousands of years. (Note: His specialty is based on the study of coal formation.)

Petrified Forest Handouts and Articles

Petrified Forest National Park: Geologic Resources Inventory Report: Natural Resource Report NPS/NRPC/GRD/NRR?2010/218 Paperback – September 17, 2013 Amazon online (3-2017). Initial publication 2010.
This paper reviews the physical resources and descriptions of them located in the Park.

www.nps.gov/pefo/planyourvisit/.
This site gives a few bits of information the PNFP feels a visitor may like to know, e.g. human history in the Park, animal/plants, activities etc.

Frequently Asked Questions https://www.nps.gov/pefo/faqs.htm
All answers are in the format of old age uniformitarian beliefs.

Petrified Forest Trees to Stone 4/2013
In this flyer the comment is made concerning fossilization, "This solution filled the cells, eventually replaced the cell walls in some cases, crystallizing as the mineral quartz **after a long time**." Author's comment: Some articles comment on thousands of years but i**t does not take a long time to fossilize wood** as has been discussed in the Geology section. Plate tectonic activity is discussed as well.

Petrified Forest Dinosaurs and Other Animals 4/2013
Invertebrates, Freshwater fish, Amphibians, Archosaurs and Therapsids are listed along with estimated description of each animal. Author's comment: though some fish can be in salt water as well, e.g. Coelacanth. I also find it a bit conflicting that there would be the assessment of fresh water fish in the setting of so many strata that are heavily "salt" or mineral laden.

Petrified Forest Messages in Stone 4/2013
It should be noted that there are similarities in Petroglyphs found among ancient peoples and modern (aboriginal). An example would be the "horned serpent" as well as some deity symbols and human forms. Per the brochure there are 1000 archaeological sites, most not accessible in the Park. But petroglyphs can be seen at the Newspaper Rock outlook. Petroglyphs are extremely difficult to date.

Petrified Forest Archaeology 4/2013
They list a Paleo-Indian culture at 13,500-8,000 BCE. This date is unlikely as there is no accurate radiometric test that can absolutely test for it. (See references on radiometric dating.) Basketmaker and Pueblo Periods are mentioned. The flyer comments, "A large percentage of the recorded sites at Petrified Forest National Park belong to Pueblo II-III. By the end of Pueblo IV, most of the Petrified Forest area appears to have been depopulated. But people still used the region for a travel corridor or for resources. They put the time periods as 500 BCE-650 CE, 950-1300 CE, and 1300-1450 CE.

Petrified Forest Common Plants 4/2013
This is an informative brochure on the plants, flowers, trees, cactuses etc. that can be grown in such an arid area.

Petrified Forest Common Animals 4/2013
A nice simple listing of various present living desert animals in the Park.

Petrified Forest Geologic Formation https://www.nps.gov/pefo/learn/nature/geologicformations.htm.
Of course there is a description of each strata/formation but it is put within the assumptions of uniformitarian philosophy.

Petrified Forest Newspaper Rock Glossy multipage handout 2016-17.
Nice pictures. The nomenclature may change.

Geologic Map of Petrified Forest National Park, Arizona J.W. Martz1, W.G. Parker1, L. Skinner, J.J. Raucci, P. Umhoefer and R.C. Blakey, 2012, published by the Arizona Geological Survey www.azgs.az.gov | repository.azgs.az.gov. http://repository.azgs.az.gov/sites/default/files/dlio/files/nid1487/cm-12-a_petrified_forest_report.pdf
This booklet gives precise locations of various strata and a description of them. Again, they are described in long ages and uniformitarian concepts of formation. Also, sand dune formation can occur underwater (see references below).

Park Boundary Expansion https://www.nps.gov/pefo/learn/news/park-boundary-expansion.htm
"The park is adding 26,093 acres of badlands, grasslands, and riparian area - very scenic with abundant wildlife in the grasslands and riparian areas and miles of cliff line with the potential for archeological sites. At least a third of the badland exposures are from the Blue Mesa Member which is rich in fossil plants and vertebrates. A good percentage is Sonsela Member with some petrified wood deposits. There are also many outcrops of the Petrified Forest Member that has not been explored but has a high fossil potential." This appears to have occurred in 2013-14.

Rocking Through The Ages: From Fossils To Petroglyphs.
https://www.nps.gov/pefo/learn/education/upload/TreeOrientation_BKLT_LEGAL.pdf
NPS handout for children. Unfortunately it teaches them the well-known scenario of long ages and local streams with occasional local floods etc. "The petrified trees you will be studying are Araucarioxylon arizonicum, a cone-bearing tree of the Triassic Period, alive over 200 million years ago. **These trees may have reached heights of over 61 meters (200 feet) and over 3 meters (9 feet) in diameter. They grew with straight trunks and had branches from the bottom to the top of the tree trunk.** This area was once **a low lying flood plain** with lots of different plants and many kinds of animal life. The petrified wood found today is only one example of ancient life found in this area."

Plant and Animal Fossils

The beginning of the Age of the Dinosaurs, 1986, Padian K. ed.; Cobert B.H. Historical Aspects of the Triassic-Jurassic Boundary Problem. Pg 9-19.
"The discussion of the Triassic-Jurrassic boundary problem as seen within the sequence of Glen Canyon formations in southwestern North America and particulary the comparisons that have been made between certain tetrapods in the **Moenave and Kayenti formations of Arizona** and the Red Beds of South Africa, leads to a consideration at this place of the **South Africa Karroo System**. The full succession, from the Dwyka Series in the base, through the overlyingEcca Series, The Beaufort Series, and, at the top, the Stormberg Series-in all comprising some 35,000 ft. of sediments occupying the Great Karroo Basin-**is one of the classic examples of continental sedimentation.**"

The beginning of the Age of the Dinosaurs, 1986, Padian K. ed.; Ash S. The Beginning of Age of Dinosaurs: Faunal Changes pg. 23-29.
"*Ginkoides.* Certain Late Triassic and Early Triassic fossils so closely resemble the leaf of the living Gingko that they are placed in that genus or in the *Ginkgoites.* During the Lower Jurassic. By this time, the order was worldwide in distribution."

The World of Coelophysis-A New Dinosaur of 200 Million Years Ago, 1981, Fisher D. ed. New York State Museum and Science Service, Circular 49 Geologic Survey December 1981.
On page 6 there are two figures: tracks found near Nyack NY and placed in the Brunswick Formation of the Newark Group in the upper Triassic/Lower Jurassic boundary. Initially the tracks were called Grallator but have now been designated Coelophysis. It is presented in many books and articles that the Newark and Chinle Formations are similar. Yet there is a huge separation between the two formations where there is no Triassic in the Midwest US; but the Chinle is in the Southwest US and the Newark along the East Coast of the US.

PFNP Geologic Resources Inventory Report (1910) gives a reported list of vertebrate discoveries in the Park as of 2010 on page 7.
These partial skeletons or fragments of Aetosaur-*Calyptosuchus*, *Rioarribasuchus*; Phytosaur-complete skeleton *Leptosuchus*; crocodylomorph-*Parrisha;* a fragment of *Trilophosaurus*; *Vancleavea*; Twelve skeletons of the Archosaur *Revuelosaurus;* lungfish; "dinosaurs"; fragments and skulls of *Metoposaurus*.

Ferns From the Chinle Formation (Upper Triassic) in the Fort Wingate Area, New Mexico Geological Survey Professional Paper 613-D, 1969, by S.R. Ash. https://pubs.usgs.gov/pp/0613d/report.pdf

D 9 "According to Kunz, volcanic action overthrew the trees in what is now eastern Arizona, and they were covered with volcanic ashes and tufa. Then hot water associated with the volcanic action and, charged with silica and other minerals, percolated through the deposit. The water infiltrated the trees, and the minerals in solution in the water slowly replaced the organic material in the wood.

D 13 "H. E. Gregory (1917, p. 49-50) **Gregory was of the opinion that all the petrified wood in the Chinle and Shinarump had been carried some distance as driftwood by flooding streams before it was buried and fossilized. His evidence included the fact that most of the logs have worn surfaces on both ends and rarely have roots or limbs attached. In addition, many of the logs appear to be battered and are usually without bark.** According to Gregory, the stumps reported to be in place by Ward (1905, p. 33-34) actually did not have roots extending downward and obviously did not grow where they are now found. He also noted that the **sandstone beds containing petrified wood are cross-bedded and lenticular, which suggests that they were deposited in a fluvial environment.**" Author's note: **These comments are similar to what I see in the Chinle Formation and argue strongly against the meandering stream going through a forest as presented at the PFNP.**

D 15 "The late **Roland W. Brown described two plant fossils from the Upper Triassic Dolores Formation** in southwestern Colorado in 1956. One fossil he referred to Brae Jiypfiyllum munsteri Schenk. **The other is a palm like leaf, which he named Sanmiguelia lewisi.** It has large elliptical, strongly pleated leaves attached by sheathing petioles to a rather stout stem. No new data bearing on the status of Sanmiguelia has been published since Brown's original study, and the fossil has not been accepted unequivocally as an angiosperm by all authorities. In a review of the evidence of pre-Cretaceous angiosperms, Scott, Barghoorn, and Leopold (1960) admit that Sanmiguelia is the best known example of a pre-Cretaceous angiosperm." **Author's note: The possibility of angiosperms in the Chinle would argue for mixing of flora during flooding and against evolutionary transitioning between the designated Triassic and Jurassic.**

D 25-27 "Representatives of most major plant groups have been described from the Chinle Formation...In his comprehensive study of the Chinle flora, Daugherty (1941) concluded that it was closely related to the flora in the Upper Triassic Newark Group of the Eastern United States. This conclusion was based on six species that he thought were common to both floras and six other species in the Chinle that he considered to be similar to forms in the Newark...The ninth species, A. *arizonicum*, may be closely related to a form in the Newark." Note: Ash did not think some of the flora in the Chinle were close to that of the Newark Formation on the East Coast of the U.S. However, many geologists still consider them of similar structural elements.

Processes of Fossilization

Rapid wood silicification in hot spring water: an explanation of silicification of wood during the Earth's history, Hisatada Akahane, T. Furuno, H. Miyajima, and S. Yamamoto, 2004, published in Sedimentary Geology Volume 169, Issues 3–4, 15 July 2004, Pages 219–228.

In this article it was noted that hot mineral springs containing a high content of silica had impregnated naturally fallen wood limbs. Under experimental conditions mimicking those of the spring a 40%, by weight, impregnation of silica into the cell lumina of modern wood tissue (alder wood *alnus pendula*) occurred within 7 years.

Soft Tissue Fossilization: Evidence for Sudden, Extensive Destruction of Life Consistent with the Genesis Flood Account, Vera Everett, 2009, published in Answers in Genesis online 4/7/17 https://answersingenesis.org/fossils/how-are-fossils-formed/soft-tissue-fossilization.

"The precipitation of these fossilizing minerals from solution will depend on factors that include temperature, ion concentration, pressure, pH, and the tendency of a particular mineral to crystallize or nucleate on a specific substance. Minerals will precipitate quickly under the right conditions. Quick changes in pressure, temperature, pH, or ion concentration can cause rapid precipitation. In order to precipitate, the mineral ions have to locally reach or be beyond saturation point. This can happen through changes in temperature, changes in mineral ion concentration, and changes in pH. The acidity can determine which mineral will precipitate. The clay mineral, illite, is stable in a very acidic environment, one in which the aragonite in shells and the apatite in bones will actually dissolve. Apatite will precipitate at a pH in which the carbonates stay in solution. As the pH gets more basic, the magnesium and calcium carbonates will precipitate…Wood can be infiltrated and permeated by water carrying silica in solution. Volcanic ash is often the source of the silica. The vacant spaces in the wood fill up with silica in a process called permineralization. This process does not take millions of years. Wood has silicified quickly in the natural environment. Pieces of fresh wood 5–8 cm in diameter and 10 cm long were placed in the Tateyama Hot Spring. The spring water, with a pH of 3, an average water temperature of 70°C, and a high silica content, precipitates amorphous opal spheres that range in size up to 10 μm in diameter. The water permeates the wood and the opal likely silicifies the wood by depositing spheres on the surfaces. After just seven years almost 40% of the wood's weight was silica."

Impact of Silicon-Based Chemicals on Selected Physical and Mechanical Properties of Wood, Vlastimil Borůvka, A. Zeidler, S. Doubek, 2016, published in WOOD RESEARCH 61 (4): 2016 513-524.

"Natural mineralization may take millions of years, but it can be successfully done faster under special conditions (Fengel 1991, Snelling 1995, Akahane et al. 2004, Kim et al. 2009). Artificial oversaturation of wood with silicates leads to structural changes similar to those in natural mineralization. The mineral solution penetrates into the wood, where it remains in lumens, or penetrates the cell walls where it can chemically react with some of the components (Hill 2006)."

Instant petrified wood. A.A. Snelling, 1995, published in Creation 17(4): 38-40.

"The chemical components used to artificially petrify wood can be found in natural settings around volcanoes and within sedimentary strata. Is it possible then that natural petrifaction can occur rapidly by these processes? Indeed! Sigleo[4] reported silica deposition rates into blocks of wood in alkaline springs at Yellowstone National Park (USA) of between 0.1 and 4.0 mm/yr."

Petrified Wood: The Silification of Wood by Permineralization, Mike Viney, www.semanticscholar.org/paper/Petrified-Wood-%3A-The-Silicification-of-Wood-by-Wood-Viney, pg. 23.

"The formation of silicified wood includes permineralization, replacement and recrystallization. When wood is permeated by silica solution, hydrogen bonding links silicic acid to cellulose making up the inner cell walls. Silicic acid is polymerized into opal through water loss. Silica that initially fixes to the wood structure is amorphous and can recrystallize to a more stable form. Layers of silica are deposited with the wood acting as a template. Cell walls may be replaced with silica as permineralization continues. Lignin is the most decay resistant compound in wood and continues to act as a template for structural detail. Studies of silicified wood from Florissant and Chemnitz demonstrate that cell walls and lumina can be preserved in different forms of silica. Thus, forms of silica that serve as the initial replicating material for cell walls and open spaces may be different from what initially permineralized the wood."

Process of Compressibility and Degradation of Wood (Non-fossilized)

The compression of wood, 1941, R. M. Seborg and A. J. Stamm, USFS. https://www.fpl.fs.fed.us/documnts/fplmisc/rpt1258.pdf

"Most of the measurements were made on 1/2 in. thick disks 4.25 in. diameter, including both the heartwood and sapwood of flat-sawed hickory. The compression of wood increased with an increase in both the moisture content of the wood, up to fiber saturation, and the pressing temperature. Wood which contains 20 percent or more moisture, when pressed at 325° F., retains at least 50 percent of its compression even after soaking in water and re-drying." They used pressures from 0-4000 psi with significant change in compression occurring at about 800-2000 psi, 26% water content, and heated at 325 degrees Fahrenheit. Compression was better retained at these higher conditions.

Method for impregnation of wood by pressure surges during immersion. US 3987219 A. (US Patent probably for possible commercial use), 1973, by E Arvidsson. https://www.google.ch/patents/US3987219

"If it is desired to impregnate wood against rot in a durable manner, every cell in the wood must be reached by the impregnation medium which makes the cell resistant to the rot fungi which otherwise would find their food there. One method of achieving this result is to let suitable salts dissolved in water be fixed in the walls of the cells. Basic inventions concerning vacuum-pressure processes for impregnation were made already more than a hundred years ago. The invention according to Swedish Pat. No. 227 747 was the first to provide an effective and inexpensive arrangement for such impregnation. It has made impregnated wood easily accessible for small consumers of such material. This invention is an improvement of the process as well as the arrangement according to Swedish Pat. No. 227 747. According to this patent the wood is submerged in an impregnation liquid and air contained in the cells is extracted by a high vacuum, which is produced by a jet of liquid pump which is connected to the distribution pump for the impregnation liquid. The intensity of the pressure surges increases successively within the said range as the cells are filled with impregnation liquid, he surges reduce the pore volume of the wood which is not occupied by the liquid and therefore at each new surge the liquid is rapidly forced further and further into the wood." This process involves pressure, removal of oxygen from the wood by vacuum, and a solvent to penetrate the wood. Salt was used over one hundred years ago per the article.

Viscoelastic thermal compression of wood. US 7404422 B2, 2008, F. A. Kamke, H. Sizemore III, 2008, https://www.google.com/patents/US7404422 (US Patent)

"CLAIMS: 1. A method for densification of wood components or composite wood, comprising the steps of i) heating and conditioning said wood components or composite wood to at least a glass transition temperature of said wood components or composite wood, then ii) inducing mechanosorption of said wood components or composite wood, then iii) compressing said wood components or composite wood to form a high density wood product, then iv) annealing said high density wood product, and then v) reducing the temperature of said high density wood product below said glass transition temperature." Note: many of the conditions for impregnation of the wood, heat/pressure/solute could have been present at the time of burial of trees and plant life in the PFNP and resulted in rapid impregnation and deformation now seen in the fossil trees there.

Disappearance and Compressibility of Buried Pine Wood in a Warm Temperate Soil Environment. 1991, H. L. Gholz, L. M. Krazynski, and B.G. Volk Ecol Appl. 1991 Feb;1(1):85-88. doi: 10.2307/1941849. https://www.ncbi.nlm.nih.gov/pubmed/27755680

"Abstract: The rate of disappearance of buried pine wood in Florida was found to be 15%/yr. As consumption by microorganisms and termites proceeded, the wood also became more compressible. After only 5 yr, consumption and compression could account for 60-70% loss of original volume of wood under pavement near the surface of an embankment. This large volume loss occurring in a relatively short time period may be responsible for many surface deformations in pavements and weaknesses in other embankments where wood may occur as a contaminant."

Radiometric Dating Methods

Creation's Tiny Mystery, Robert Gentry, 1986, published by Earth Science Associates, Knoxville, Tennessee, pg. 51-56. On page 55 he stated regarding polonium haloes:

"**Their occurrence in three geological formations suggests that they all originated at about the same time in agreement with a flood-related scenario.**" The Three geological formations referred to are the Triassic, Jurassic, and Eocene on the Colorado Plateau.

Response to Geochronology: Understanding the Uncertainties, a presentation by Dr Justin Payne http://creation.com/geochronology-uncertainties.

"What is often overlooked is that, underlying the assumptions, there are presuppositions that can have a profound impact on the assumptions and, therefore, the analyses and conclusions, even to the extent of precluding getting the correct answer. Dr Payne alludes to this near the end of his talk when he says something to effect of "people who believe in Flood Geology are forced, implying that the acceptance of Flood Geology as a presupposition would have an impact on—indeed 'force'—a particular interpretation of the data."

"**The fundamental presupposition of most scientists these days is that of naturalism, the view that nature is all that there is.** Sometimes this is referred to as materialism, the view that matter is all that there is. By definition, these exclude the potential existence of, respectively, the supernatural and the immaterial. A consequence of this is that **it forces an explanation of everything, including the origin of things,** using only matter and observed natural processes—in other words, the laws of physics and chemistry." (bold added)

Thousands Not Billions, Donald Young, 2005, published by Mast Books Green Forest Arkansas.
Pg. 174-179 summarized the finding of the RATE (Radio-Isotopes and the Age of the Earth) study.
1) "For some years there has been the growing realization that **C-14 atoms are found where they are not expected.** With a half-life of 5,300 years C-14 should no longer exist within "ancient' fossils, carbonates or coal." It has been found in every coal sample and even diamonds tested per Dr Young. **2)** "**Zircons** play a prominent part in the RATE studies." These crystals are found in deep granites (e.g. assumed "old") and **still contain helium which should have long ago escaped. 3)** "**Radio halos** are tiny spherical defects in rocks present in abundance in **granites**. (Some) appear to have **formed during the decay of short-lived polonium**." Note: this observation has been used for the view regarding the rapid formation of the granites (see Gentry). Parentheses by author. **4)** "Many rock units worldwide were analyzed by radioisotope dating techniques." The studies were compiled for many **parent-daughter isotopes** and the results gave "examples of **concordance, discordance or disagreement**. Both extremes often occurred for the same rock unit. The conclusion is that **no isochron can be trusted with confidence**." **5)** "There are there **important assumptions made in radioisotope dating** (and) **found subject to failure**."

First assumption: that "initial conditions of rock sample can be determined." Too many discordant dates in isochrons found on sampling. Second assumption: "that the open or closed nature of rock samples can be determined and quantified." He mentions the mixing of mantle and "crustal isotopes" with the rock samples. Polonium radiohalos reveal hydrothermal movement through the rocks. Third assumption: "nuclear half-lives have remained constant throughout history. He uses helium still present in zircon crystals to argue against this assumption. He also goes on to discuss accelerated nuclear decay as a possibility in the Creation Week and the Noah Flood. He also lists linguistic analyses in support of a short age for the earth.

Underwater Sand Dunes

https://answersingenesis.org/geology/grand-canyon-facts/startling-evidence-for-noahs-flood/
Originally in Creation *15, no 1 (December 1992): 46-50. Startling Evidence for Noah's Flood Footprints and Sand 'Dunes' in a Grand Canyon Sandstone!* by Dr. Andrew A. Snelling and Dr. Steve Austin
"The footprint track ways in the Coconino Sandstone have recently been re-examined in the light of experimental studies by Dr Leonard Brand of Loma Linda University in California. A **detailed statistical analysis of these data led to the conclusion, with a high degree of probability that the fossil tracks must have been made underwater**. The desert sand dune model for the origin of the Coconino Sandstone has also recently been challenged by Glen Visher, Professor of Geology at the University of Tulsa in Oklahoma, and not a creationist geologist. Visher noted that **large storms, or amplified tides, today produce submarine sand dunes called 'sand waves'.** These modern sand waves on the sea floor contain large cross beds composed of sand with very high quartz purity. **Visher has thus interpreted the Coconino Sandstone as a submarine sand wave deposit accumulated by water, not wind.**"

ABOUT THE AUTHOR

Dr. John G. Leslie received a PhD in 2012 in the field of Archaeology and Biblical History from Trinity Southwest University (TSWU), Albuquerque, NM. When he took part in the annual TSWU dig of a biblical site in the Jordan Valley, he studied the human and animal bone artifacts.

Dr Leslie has also been practicing medicine in a medically underserved area of the Southwest United States for over 25 years. He made a commitment to God to do this after he returned to medical school in his mid-thirties. The training for an MD was at Oral Roberts University, with graduation in 1989. He then went to the University of Oklahoma Tulsa Branch to complete a dual residency in Internal Medicine and Pediatrics in 1993, and subsequently became board certified in both fields.

Prior to going to medical school Dr Leslie completed a PhD in Experimental Pathology in 1980 from the University of Utah. He worked with one of the world's leading specialists studying the structure and metabolism of Elastin, one of the key structural proteins of the body. After completing this degree he worked at Monash University in Melbourne, Australia as a post-doctoral researcher for 4 years till returning to medical school. His undergraduate BS degree was from the American University in Washington, DC in 1973.[1]

Dr Leslie committed his life to Jesus Christ, and experienced being "born again" late 1974. Several people were key to this happening in his life and the most significant was his godly mother Doris Gregory Leslie. He and his wife Barbara have served God in various ways during their married life of 39 years. They have three married children, and numerous grandchildren.

You can visit the author's website at **www.DefendingTheChristianFaith.org**

To God be the Glory.

Made in the USA
Monee, IL
05 November 2026

67386675R00064